T0257530

Handbook of Expert Systems

Handbook of Expert Systems

Edited by **Mick Benson**

CLANRYE
INTERNATIONAL

New Jersey

Published by Clanrye International,
55 Van Reypen Street,
Jersey City, NJ 07306, USA
www.clanryeinternational.com

Handbook of Expert Systems
Edited by Mick Benson

International Standard Book Number: 978-1-63240-271-4 (Hardback)

Contents

Preface

The aim of this book is to educate the readers regarding expert systems. Intelligent systems aka expert systems fulfill the purpose of their name in the way that they act as facilitators in this procedure. These are systems that are based on specialist information on any topic in order to emulate human capability in the exact field. To attain this information, the knowledge experts, also called software engineers, require expanding of methodologies for intelligent systems. In this field there is still no unified methodology that supplies efficient techniques and tools to facilitate any new advancement.

This book is a result of research of several months to collate the most relevant data in the field.

When I was approached with the idea of this book and the proposal to edit it, I was overwhelmed. It gave me an opportunity to reach out to all those who share a common interest with me in this field. I had 3 main parameters for editing this text:

1. Accuracy – The data and information provided in this book should be up-to-date and valuable to the readers.

2. Structure – The data must be presented in a structured format for easy understanding and better grasping of the readers.

3. Universal Approach – This book not only targets students but also experts and innovators in the field, thus my aim was to present topics which are of use to all.

Thus, it took me a couple of months to finish the editing of this book.

I would like to make a special mention of my publisher who considered me worthy of this opportunity and also supported me throughout the editing process. I would also like to thank the editing team at the back-end who extended their help whenever required.

Editor

Industrial Applications

Electric Power System Operation Decision Support by Expert System Built with Paraconsistent Annotated Logic

João Inácio Da Silva Filho, Alexandre Shozo Onuki,
Luís Fernando Pompeo Ferrara,
Maurício Conceição Mário, José de Melo Camargo,
Dorotéa Vilanova Garcia,
Marcos Rosa dos Santos and Alexandre Rocco

Additional information is available at the end of the chapter

1. Introduction

In the last decades there has been a gradual increase of the industrial park in several countries which demands energy, especially electric power. This situation caused a considerable expansion of the sector responsible for generation and distribution of electric power in a very short period of time. The rapid expansion caused a significant increase of generation sources and of distribution branches of electric power which originated enormous and complex agglomerates with interconnections among themselves and with a certain degree of dependence and vulnerability.

This complexity of electric power systems brought up technical needs and technological challenges in order to obtain efficient methods to monitor the variables of the electric quantities which express the operational normality state of the networks. Together with the need of energy production the priority, the sentiment of priority given to the extraction forms and the correct usage of energy by mankind came up. This sentiment brought up new public policies of generation and distribution of electric power. Currently, the laws concerning this issue have concentrated on the supervision of the concessionary companies which are responsible for the provision of electric power, making sure that these companies offer energy with a high-quality level to the consumers. Besides these obligations, the utilities companies also have the need of markets and because of that they have big interest in the modernization of the management, monitoring and control of their power systems. Due to these facts, a huge effort and investment of the concessionary companies in the research are which deals with the quality of electric energy offered to consumers has been verified [1].

One of the indexes which measure the quality of electric power offered to consumers is obtained by the number of outages or failures in the energy distribution of the electric power system (and time length) in a certain period of time. When an electric power system is overloaded, it is possible that its equipments may be disconnected by relays which act as protection, causing the interruption of electric power in certain areas covered by its transmission networks.

That being so, it is very important to research new methods that effectively evaluate the state of outage risk due to overloads in the system because it is extremely important, in order to decrease indexes of energy interruption to consumers, to manage the power loads with permanent monitoring of the distribution lines of electric power. However, in the case of an interruption, it is also fundamental to make decisions quickly and safely in order to reconnect systems after an outage.

1.1. Electric Power System Overview

A typical electric power system can be divided in generation and systems of transmission, sub-transmission and distribution. The transmission system interconnects generating stations to large substations located near load centers generally using aerial electric transmission lines. The sub-transmission system distributes energy to an entire district and usually uses aerial electric transmission lines. The distribution system transports energy from the local substation to individual houses, using aerial or underground transmission lines [7]. A typical electric power system can be seen in Figure 1.

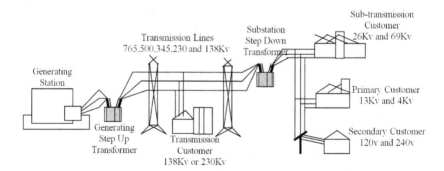

Figure 1. Simplified picture of a typical electric power system.

A typical transmission system has three phase conductors to take the electric current and transport power. Each phase of the transmission line is built with two, three or four parallel conductors separated approximately by 1.5 ft (0.5 m) [3][7].

1.2. Main Problems of an Electric Power System

It is well known that operation failures in an electric power system are unavoidable and there are a large number of reasons why these interruptions happen. This situation is due to the natural conditions of an electric power system in which failures may happen because of internal or external causes, such as consequences of environmental physical phenomena which are beyond the physical specifications of the electric systems or even human error [3].

There also is a fundamental limitation on the electricity distribution: with few exceptions, electric power cannot be stored which means that it must be generated as it is demanded. That being so, an electric power system must provide electric power with safety and with acceptable tolerance ranges either for a normal load or for a demand condition of maximum load or of peak [1][2]. Since the demand periods of peak load, due to the several types of industry or to different types of housing, are different from region to region, this natural condition of electricity brings up a problem of generation control and transmission.

In certain regions industries can be more productive in certain times of the day and show a drop in demand because of lunch time, so the demanded energy has variations during the day. In highly densely populated regions where there is a lot of night life businesses, the energy demand is larger in the evening and also depends on the day of the week and even on the season of the year.

Besides this problem particular to each industrial park or city, another factor to consider is the climate of the region where these industries or residences are located. In regions with very hot weather, turning on air conditioning system in the hottest period of the year, the energy demand values are higher in the late afternoon. For regions with very cold weather the energy demand has different effects on the global load of the electric power system because in the coldest period of the year heaters are turned on in the morning and in the evening.

These situations of difficult control makes power or demand for electric energy vary and besides, the capacity of a transmission line to transport electric power is limited by physical and electrical parameters of its conductors. In order to avoid interruptions these conductors of electric energy in any load conditions must be sufficient to respond to the demand within limits such that their safety relays will not be activated.

Transmission lines are subject to environmental adversities including large temperature variations, high winds, storms, etc. Thunders that fall on transmission towers cause high voltages and propagating waves in the transmission lines which usually cause the destruction of isolators and as a consequence of that the protection relays interrupt the power transmission through the networks [7].

1.3. The voltage variation and Overcurrent as Overload Risk Factors

According to what was seen, in an electric power system the loads represented by the electric power consumers, such as electric machines of industries, lighting systems, heating devices of residences and refrigeration systems of businesses, are not static. They are constantly changing, being turned on and off with value variations which may lead to overload. The overloads are outage risks for the whole system because it increases the intensity of the electric current (overcurrent) in the lines and can heat the conductors, increasing their temperature and causing permanent damage with the interruption of energy transmission.

The existence of load variations requires precise equipments which adjust the voltage in the line, because the overload causes the voltage outage. The voltage variation can aggravate the electric power system state with emergence of large intensities in distribution branches. Because of that, the decrease value or the voltage outage (under-voltage) and the intensity value of the electric current (overcurrent) are two important risk factors to the monitoring of transmission lines of electric power. That being so, the monitoring of the ranges of voltage outage and of maximum current of an electric power system are used as a diagnosis of overloads. In order to increase the quality index these two factors must be constantly monitored because if they both are out of the ranges specified in their projects the possibility of disconnection of the electric power system will be higher.

2. Artificial Intelligence and Electric Power Systems

Artificial intelligence techniques have become necessary to procedures of monitoring, management and control of electric power systems. The current expansion of electric power systems is physically verified by the increase of branches and by the way distribution lines are installed: generators and loads are interconnected with the distribution lines through multiple paths (radial form) and in ring among them. This technique increases the confidence index on the system because the failure of one line does not cause a total failure of the system and can provide the transmission of electric power from other of its branches. Every new technique offers certain advantages, however when a new technique is implemented, there is an increase of the complexity of the electric power system. Because of that there is the need of an efficient protection in which the sector responsible for the energy transmission and distribution as well as the generating sector are controlled in a quick and efficient way in order to keep the power generated according to the charges required. That said so, energy generation must be kept according to the conditions established by the load and comply with the conditions in which the protection systems are capable of prevent failures of the generation equipment due to possible overloads [3][7].

Recently, new techniques from artificial intelligence (AI) made possible to connect multiple generating sources of electric power as well as loads to the transmission system. However, although all these new factors make access easier, they may cause problems by destabilizing the system, which requires a sophisticated AI based control to assure capable and efficient control of the generation according to the demand. These actions which modify the operational state of the electric power systems at any time must be controlled in order to provide power transference in a safe and coordinated way.

Nowadays, the software SCADA is installed in the electric power systems. SCADA stands for Supervisory Control and Data Acquisition and provides in real-time large amounts of information about values related to electric quantities of the electric power system which can be obtained from points of the transmission network. Such quantities are, for example, voltage, current and potency. The information obtained by the SCADA system are used as input for the analysis systems and decision-making [1][8][9]. However, due to the large amount of information received by the control centers, interference and natural failures in the synchronization of the transmission, it is not always that the information brought for analysis by SCADA are complete. Because of that, a databank is created and contains ambiguous, vague and contradicting data. Due to the nature of these data a human operator could be led to make false interpretations or even make wrong decisions which could lead to huge losses and delay of the system restoring [1] with damage to the quality index desired. That being so, it became very urgent to have computational treatment with expert systems dedicated to interpretation, data analysis and the presentation of suggestions as a way of supporting the operation of electric power systems.

2.1. Expert Systems structured in non-classical logics

Artificial Intelligence research oriented to electric power systems has as its goal to find ways of designing new computational tools for the support of decision-making of the team responsible for the correct operational action. However, due to the large number of electric keys (breakers that modify the system's topology), to variations on the values of loads and to other several factors inherent to an electric power system, there are many difficulties to find efficient ways. The methods which use conventional binary classical logics to analyze data from the electric power system with the capability of offering suggestions for the optimized restoring after a failure has not provide good results. One of the difficulties found in the design of models based in classic logics is its condition of being defined by rigid binary laws which lead to equations which are extremely complex to reproduce models. Besides, these equations almost always lead to a combinatorial explosion. Due to this aspect, in this area of artificial intelligence, projects designed with the goal of analysis and decision-making based in classic logics has found many difficulties. It is verified that the low efficiency showed by these projects which use classic logics comes up when a large amount of data has to be computed. These data almost always have redundancies which bring up incompleteness and contradiction invalidating important information for the analysis. Some classic works use complex algorithms with good results but the computation time is very high making the response time long which is unfeasible in real conditions where an electric power system always demands quick and direct actions in order to avoid bigger damages.

2.2. Expert Systems Based on Paraconsistent logics

These problems found in classical projects lead to the conclusion that the algorithms based on the concepts of non-classic logics may show a better efficiency in the design of expert systems dedicated to the analysis and treatment of uncertain data originated in complex data-banks such as the one of an electric power system.

Based on these considerations we introduce in this paper a paraconsistent expert system (PES) with algorithms based on the theoretical concepts of the Paraconsistent logics (PL) which is a non-classic logics whose main basic theoretical features is, under certain conditions, to accept contradicting values so that the conflict does not invalidate the conclusions [5][9][10].

The paraconsistent expert system (PES) introduced in this work has the role of performing the analysis of the information coming from the electric power system in the sector of energy sub-transmission treating possible contradictions in the information signals. Through the analysis of values of voltage and electric current and the consideration of the states of connection or disconnection of the electric keys in the substations, PES informs about the risk conditions of overload and about the different configuration topologies of the electric power system.

When a failure, that triggers the interruption of the transmission of energy, happens, PES informs the operators of the sub-transmission system how to proceed with the actions for the restoring in an optimized way. Given the real-time monitoring, the analysis that PES performs is based on data before the occurrence of the failure, which allows PES to indicate the best and most efficient sequence of connecting electric keys in the interrupted section. The actions indicated by PES take into consideration the restrictions of load, technical and safety norms due to the conditions imposed to that particular situation.

3. Paraconsistent Logics – Equations and Algorithms

Aristotelian or classic logics are called so due to its origin being attributed to Aristotle and his disciples, and its foundations are supported by strict binary principles which can be concisely described by: principle of identity, principle of bivalency, principle of non-contradiction and principle of the excluded middle. Basically, all current technology is built based on the principles of the classic logics. However, due to its binary foundations, it cannot be applied or cannot offer satisfactory responses in some real situations such those where incompleteness and contradiction are expressed.

In order to overcome these difficulties and fulfill the need of satisfactorily model certain conditions of the real world, several logics, which reject some of the classic principles or which accept certain conditions not included in the classic logics, have appeared recently. The special logics are called non-classic and among them there is the paraconsistent logic (PL) which has the main property of being capable of accepting contradiction in its foundations.

3.1. Paraconsistent Annotated logics - PAL

Among the several families of paraconsistent logics there is the logics called paraconsistent annotated logics (PAL) [5] which belongs to the class of evidential logics and allows analysis of signals represented by annotations [5][9][11]. In its representation each annotation μ belongs to a finite lattice τ which assigns values to its corresponding propositional formula P.

For the PAL each evidence degree μ from its representing lattice, whose value varies from 0 and 1 in a closed interval of real numbers, assigned to the proposition P a logical state represented on the vertexes. By means of a special logical operator, the interpretations on the lattice of the PAL allow the creation of equations which provide algorithms for the paraconsistent analysis with evidence degrees extracted from real physical systems.

3.2. Paraconsistent Annotated logics with Annotation of two values - PAL2v

The paraconsistent annotated logics with annotation of two values (PAL2v) is an extension of the PAL and to each propositional formula P is assigned an annotation given by two evidence degrees as follows:

An evidence degree (μ) which is favorable to proposition P and an evidence degree (λ) which is unfavorable to proposition P.

The annotation composed by two evidence degrees (μ,λ) gives proposition P a connotation of paraconsistent logical state ε_τ which can be identified on the extreme vertices of the lattice: inconsistent ($\top$), true (t), false (F) or indeterminate ($\perp$) [9]. That being so, in the representation of the PAL2v, a paraconsistent logical signal is represented by proposition P and its annotation, which is composed by two evidence degrees, such that: $P_{(\mu,\lambda)}$ with $\mu, \lambda \subset [0,1] \in \mathfrak{R}$.

3.3. The Equations of PAL2v

The PAL2v can be studied with the unitary square of the Cartesian plane (USCP) as shown in Figure 2 where, through linear transformations, values on the two representing axes of a lattice similar to the one associated with the PAL2v.

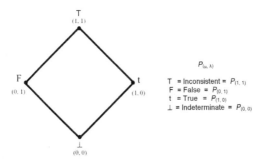

Figure 2. Lattice of four Vertexes.

Doing so, we can write paraconsistent equations on the lattice in which terminologies and conventions are established [5] around paraconsistent logical states attributed to proposition P. After the expansion actions with intensity $\sqrt{2}$, rotation of 45° with respect to the origin and translation along the vertical axis, the linear transformation is defined by:

$$T(X,Y)=(x-y, \quad x+y-1) \tag{1}$$

According to the language of the PAL2v we have:

$x = \mu$ is the Favorable evidence Degree

$y = \lambda$ is the Unfavorable evidence Degree.

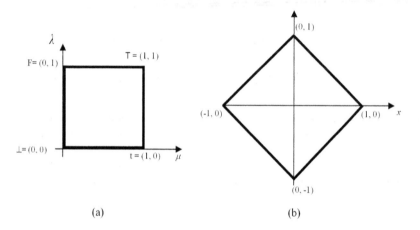

(a)   (b)

Figure 3. a) Unitary Square in the Cartesian Plane (USCP). (b) Lattice κ with another system of coordinates with values.

The first coordinate of the transformation (1) is called *Certainty Degree* D_C.

$$D_C = \mu - \lambda \tag{2}$$

The first coordinate is a real number in the closed interval [-1,+1]. The x-axis is called *"axis of the certainty degrees"*.

The second coordinate of the transformation (1) is called *Contradiction Degree* D_{ct}.

$$D_{ct} = \mu + \lambda - 1 \tag{3}$$

The second coordinate is a real number in the closed interval [-1,+1]. The y-axis is called *"axis of the contradiction degrees"*.

3.4. The Paraconsistent States Logic ε_τ

Since the linear transformation $T(X,Y)$ shown in (1) is expressed with evidence Degrees μ and λ, from (2), (3) and (1) we can represent a Paraconsistent logical state ε_τ into Lattice τ of the PAL2v [3], such that:

$$\varepsilon_{\tau(\mu,\lambda)} = (\mu - \lambda, \mu + \lambda - 1) \tag{4}$$

or

$$\varepsilon_{\tau(\mu,\lambda)} = (D_C, D_{ct}) \tag{5}$$

where: ε_τ is the Paraconsistent logical state.

D_C is the Certainty Degree obtained from the evidence Degrees μ and λ.

D_{ct} is the Contradiction Degree obtained from the evidence Degrees μ and λ.

Since the Paraconsistent logical state ε_τ can be anywhere in the lattice τ, the real Certainty Degree D_{CR} can be obtained as follows:
For $D_C > 0$ we compute:

$$D_{CR} = 1 - \sqrt{(1 - |D_C|)^2 + D_{ct}^2} \tag{6}$$

For $D_C < 0$ we compute:

$$D_{CR} = \sqrt{(1 - |D_C|)^2 + D_{ct}^2} - 1 \tag{7}$$

where: $D_C = f(\mu,\lambda)$ and $D_{ct} = f(\mu,\lambda)$

For $D_C = 0$ we consider the undefined Paraconsistent logical state with: $D_{CR} = 0$.

We compute the resulting evidence Degree which expresses the intensity of the Paraconsistent logical state ε_τ by:

$$\mu_{ER(\mu,\lambda)} = \frac{D_{CR} + 1}{2} \tag{8}$$

where:

$\mu_{ER(\mu,\lambda)}$ is the resulting evidence Degree in function of μ and λ.

D_{CR} is the real Certainty Degree (6) or (7).

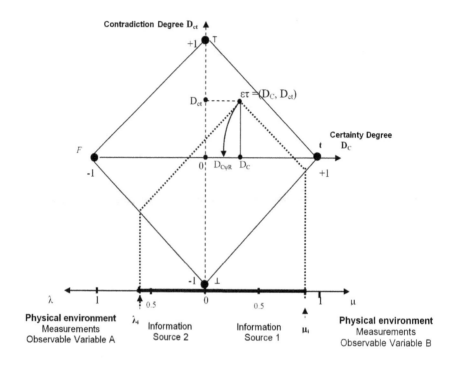

Figure 4. Representation of a Paraconsistent Logical State in to Lattice.

3.5. Algorithms of the Paraconsistent Logics

With the considerations here presented we can compute values using the equations obtained from the analysis and interpretations of the paraconsistent logics PAL2v where a paraconsistent analysis system receives information signals in the form of values of evidence degrees which vary from 0 to 1.

Through the algorithms, a paraconsistent analysis system can be built and it is capable of offering a satisfactory response from information extracted from the databank of uncertain knowledge. In this work we use 3 types of algorithms based on the PAL2v according to the following descriptions.

3.5.1. Evidence Degree Extracting Algorithm

The paraconsistent system for treatment of uncertainties may be used in many fields of knowledge where incomplete or contradictory information will receive an adequate treatment through the equations of the PAL2v. For this, the signals which will represent the evidence in relation to the proposition in analyses must be normalized and all the processing will be done in real closed interval between 0 and 1 [9].

This process for modelling the evidence degrees with linear variation can be made in its simpler form with the algorithm that will be described in the next section [13-15].

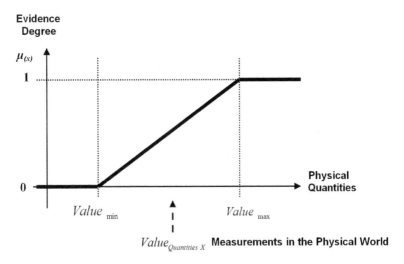

Figure 5. Graphical representation of the extraction of the Evidence Degree Algorithm - with characteristics of directly proportional variation.

3.5.2. Algorithm for Modelling/extraction of Evidence Degrees (Inputs of the PAL2v Algorithm)

3.5.2.1. Present the Maximum boundary-value to form the Discourse Universe.

$$Value_{max} = \text{....................} \tag{9}$$

3.5.2.2. Present the Minimum boundary-value to form the Discourse Universe.

$$Value_{min} = \text{....................} \tag{10}$$

3.5.2.3. Present the Value Measured of the Physical Quantities.

$$Value_{Quantities\,X} = \text{........................} \tag{11}$$

Obs: In real Physical System this value is obtained from measurements in sources of information.

3.5.2.4. Calculate the Favorable Evidence Degree μ through the equations:

$$\mu_{(x)} = \begin{cases} \dfrac{Value_{Quantities\ X} - Value_{min}}{Value_{max} - Value_{min}} \\ \quad if\ \ Value_{Quantities\ X} \in [Value_{min}, Value_{max}] \\ 1 \qquad if\ \ Value_{Quantities\ X} \geq Value_{max} \\ 0 \qquad if\ \ Value_{Quantities\ X} \leq Value_{min} \end{cases} \tag{12}$$

3.5.2.5. Calculate the Unfavorable Evidence Degree λ by Complement of the Favorable Evidence Degree.

$$\lambda_{(x)} = 1 - \mu_{(x)} \tag{13}$$

3.5.2.6. Provide the outputs.

For information source 1: Do: $\mu = \mu_{(x)}$

For information source 2: Do: $\lambda = \lambda_{(x)}$

3.5.2.7. End.

Depending on the proposition to be analyzed and on the physical properties of the quantities from which the evidences are extracted, the variation between the maximum and minimum values at the extraction of the evidence degrees can be different such as: linear and inversely proportional characteristic, exponential characteristic, logarithmic characteristic, etc. In these cases, the equations of item 3.5.2.4 are modified according to the mathematical equation of the variables which express the characteristic line or curves used in the discourse universe.

3.5.3. Algorithm of paraconsistent analysis

The main PAL2v Algorithm used in paraconsistent analyses is the PAN- Paraconsistent Analyzer Node. In an Intelligent system that works with Paraconsistent Logic some PANs are linked forming uncertainty analysis networks (PANnet) for signal information treatments [14][15][16].

3.5.3.1. Paraconsistent Analysis Node - PAN

The element capable of treating a signal that is composed of one degree of favorable evidence and another of unfavorable evidence (μ_{1a}, μ_{2a}), and provide in its output a Resulting Evidence Degree, is called basic Paraconsistent Analysis Node (PANb).

Figure 6(b) shows the representation of a PANb with two inputs of evidence degree:

μ_1 = favorable Evidence Degree of information source 1.

λ = unfavorable Evidence Degree, where: $\lambda = 1 - \mu_2$

μ_2 is a favorable Evidence Degree of information source 2.

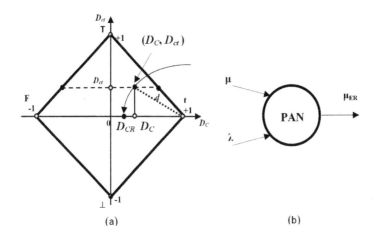

(a)   (b)

Figure 6. Finite Lattice of PAL2v and Symbol of the Paraconsistent Analyzer Node - PAN.

A lattice description uses the values obtained by the equation results in the Paraconsistent Analyzer Node Algorithm [3][13][14] that can be written in a reduced form, as follows:

1. Enter with the input values.

μ */ favorable evidence Degree $0 \le \mu \le 1$

λ */ unfavorable evidence Degree $0 \le \lambda \le 1$

2. Calculate the Contradiction Degree.

$$D_{ct} = \mu + \lambda - 1 \tag{14}$$

3. Calculate the Certainty Degree.

$$D_C = \mu - \lambda \tag{15}$$

4. Calculate the distance d of the Paraconsistent logical state into Lattice.

$$d = \sqrt{(1 - |D_C|)^2 + D_{ct}^2} \tag{16}$$

5. Compute the output signal.

If $d{\ge}1$ Then do S1= 0.5: Indefinite logical state and go to the steep 10

Or else go to the next step

6. *Calculate the real Certainty Degree.*

If $D_C{>}0$ $D_{CR}=(1-d)$

If $D_C{<}0$ $D_{CR}=(d-1)$

7. *Present the output.*

Do   S1 = D_{CR}

8. *Calculate the real Evidence Degree.*

$$\mu_{ER} = \frac{D_{CR}+1}{2} \tag{17}$$

9. *Present the output.*

Do   S1 = μ_{ER} and   S2= D_{ct}

10. *End.*

The Systems with the Paraconsistent Analysis Nodes (PAN) deal with the received signals through algorithms, and present the signals with a real evidence Degree value in the output [3].

3.5.4. Paraconsistent Algorithm Extractor of Contradiction Effects – ParaExtr$_{ctr}$

The Paraconsistent Algorithm Extractor of Contradiction effects (*ParaExtr $_{ctr}$*) is composed by connections among PANs. This configuration forms a Paraconsistent Analyze Network capable to extract the effects of the contradiction in gradual way of the signals of information that come from Uncertain Knowledge Database.

The hypothesis of extraction of the effects of the contradiction has as principle that; if the first treated signals are the most contradictory and then the result of the paraconsistent analysis will converge for a consensual value.

In his typical operation the *ParaExtr $_{ctr}$* receives a group of signals of information represented by Degrees of Evidence (μ_E) the regarding certain proposition P and, independently of other external information, it makes paraconsistent analysis in their values where, gradually, it is going extracting the effects from the contradiction to remain as output a single resulting Real Evidence Degree μ_{ER}. The μ_{ER} is the representative value of the group of input signals after the process of extraction of the effects of the contradiction.

The figure 7 shows the representation of the algorithm Extractor of Contradiction effects that uses a network of three PANs.

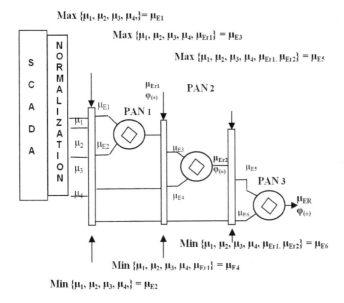

Figure 7. Paraconsistent Algorithm Extractor of Contradiction effects (ParaExtr$_{ctr}$).

The description of the *ParaExtr $_{ctr}$ Algorithm* is shown to proceed.

1. Present n values of Evidence Degrees that it composes the group in study.

$G\mu = (\mu_A, \mu_B, \mu_{C,...,} \mu_n)$ */Evidence Degrees $0 \leq \mu \leq 1$*/

2. Select the largest value among the Evidence Degrees of the group in study.

$\mu_{maxA} = max\ (\mu_A, \mu_B, \mu_{C,...,} \mu_n)$

3. Consider the largest value among the Evidence Degrees of the group in study in favorable Evidence Degree.

$\mu_{maxA} = \mu_{sel}$

4. Consider the smallest value among the Evidence Degrees of the group in study in favorable Evidence Degree.

$\mu_{minA} = min\ (\mu_A, \mu_B, \mu_{C,...,} \mu_n)$

5. Transform the smallest value among the Evidence Degrees of the group in study in unfavorable Evidence Degree.

$1 - \mu_{minA} = \lambda_{sel}$

6. Make the Paraconsistent analysis among the selected values:

$\mu_{R1} = \mu_{sel} \Diamond \lambda_{sel}$ */ where $\Diamond$ is a paraconsistent action of the PAN */

7. Increase the obtained value μ_{R1} in the group in study, excluding of this the two values μ_{max} and μ_{min}, selected previously.

$G\mu = (\mu_A, \mu_B, \mu_{C,...,} \mu_n, \mu_{R1}) - (\mu_{maxA}, \mu_{minA})$

8. Return to the item 2 until that the Group in study has only 1 element resulting from the analyses.

Go to item 2 until $G\mu = (\mu_{ER})$

4. The Paraconsistent Logical Model of The Expert System (PES$_{PAL2v}$)

An expert system is designed and developed to attend a certain and limited application of human knowledge. Moreover, equipped with an information base, it must be capable of providing a decision based on justified knowledge. Doing so, the algorithms which compose the computational programs of the expert system need to represent knowledge from the domain they have to analyze and assist the user in solving problems.

The precision of the results depends on the capability of knowledge acquisition and transference methods of this information through a computational language which can be accordingly treated and on returning a consistent response.

Following this model, the application of the paraconsistent logics PAL2v in the analysis of electric power systems is done with the reception of data corresponding to the values of voltage and current captured by the SCADA system where they are normalized in order to be adjusted to the concepts of the PAL2v. These signals receive adequate treatments by the PAN algorithms in their normal configuration or interconnected, composing networks of blocks which extract the contradiction effects building a paraconsistent logical model related to the risk state of overload on the system.

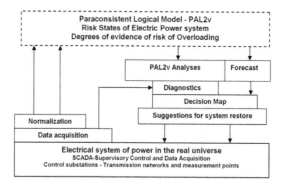

Figure 8. Paraconsistent logical model composed by risk evidence degrees obtained with values of current and voltage captures in the real electric power system.

According to the paraconsistent expert system (PES$_{PAL2v}$) the real electric power system in operation owns its paraconsistent logical model based on evidence degrees whose propositions are related to states of outage risks by overloading.

Figure 8 shows the paraconsistent logical model composed by risk evidence degrees configured by the real electric power system.

4.1. Contingency Analysis for Electric Power Systems Using Paraconsistent Logics

The operation of the PES$_{PAL2v}$ starts when there is the occurrence of a contingency or failures with electric power outage. This is when the algorithms of the Paraconsistent Expert System receive data for analysis of pre-failure states which were stored in the SCADA system database. This allows the PES$_{PAL2v}$ to check the risk degrees of overloading with measures of voltage and current before the occurrence. The verification of the resultant evidence degrees detects with a certain evidence degree which branch of the power network had a high overloading degree risk before the occurrence.

This pre-failure analysis offers conditions such that at the time of contingency we can compare the obtained evidence degree of overloading risk with the risk state that the system had in the condition previous to the event. So, it is possible, through the results from the comparative analysis between the two moments and the condition of the topology of the electric network in its area affected by the contingency that the PES$_{LPA2v}$ can do the most convenient adaptation of maneuvers to be applied to the optimized restoring of the electric power system.

According to the results of the comparisons among the evidence degrees of overloading risk, the analysis of the paraconsistent expert system PES$_{PAL2v}$ will suggest control actions to the restoring of the electric power system based in three states of the sub-transmission system [13].

These analysis procedures can be seen on Figure 9.

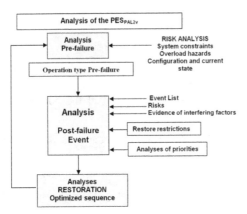

Figure 9. Flowchart of the analysis states in the process.

1. Pre-failure – consists in the analysis of the sub-transmission system in operation.

2. Post-failure – consists in the analysis of the sub-transmission system at the instant of the contingency.

3. Restoring – consists in the analysis of the sub-transmission system after the contingency.

4.1.1. Propositions Used in the Paraconsistent analysis

The paraconsistent analysis in the PES_{PAL2v} is based on the configurations of the PANs where the paraconsistent logical signals are extracted from measured values of voltage and electric current. The PAL2v analysis is performed with applied paraconsistent logical signals with annotations composed of evidence degrees related to 5 partial propositions.

The two first analyze the tension outage and overcurrent at the measurement points and generate evidence degrees related to the existence of overloading in the sub-transmission network. They are:

Pp1: There is overcurrent in the electric power network

Pp2: There is sub-voltage in the electric power network

Next, through the PANs algorithms, the paraconsistent analysis with the degrees of sub-voltage and overcurrent generated by this initial analysis which result evidence degrees, now related to the annotation of the object proposition:

Po: There is the risk of drop by overload in the electric power network.

For the decision-making about the optimized restoring of the sub-transmission system after a contingency, PES_{PAL2v} still analyzes other two propositions related to the restrictions and the topology of the power network:

Po1: There are restrictions of loads in the electric power network.

Po2: The network topology is ideal for the current situation.

That being so, the sequence of maneuvers which are offered to the operation will be conditioned directly to the configuration of topologies, technical norms and restrictions which involve the area of the sub-transmission system of the power network affected by the contingency.

The classification performed by the paraconsistent analysis network (PANet) generates a resulting evidence signal whose value will define the type of operation and sequences of restoring closest to the ideal, given the conditions of the sub-transmission system.

Figure 10 shows the pre-failure analysis with its partial propositions which generate the evidence degrees for its object proposition and whose result will be used for the post-failure analysis.

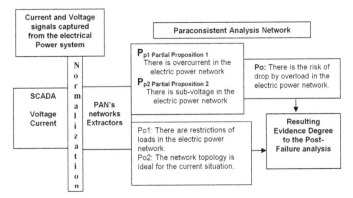

Figure 10. Representation of the pre-failure analysis with its partial prepositions which generate evidence degrees for the post-failure analysis.

5. Project of the Paraconsistent Expert System - PES$_{PAL2v}$

The project of PES$_{PAL2v}$ starts with the definition of the methods of acquisition of data through the evidence degree extracting algorithms with the goal of generating the paraconsistent logical signals for the analysis network composed of algorithms based on the PAL2v.

5.1. Acquisition of Measurements

The first task to be performed by the paraconsistent expert system PES$_{PAL2v}$ is the acquisition of values of measurements performed in the system so that the overload risk levels can be detected. For this purpose, we use the data available in the SCADA (Supervisory Control and Data Acquisition) system which, in this phase, has to receive several types of measurements from the field.

The SCADA system is responsible for the interface between the measurements of electric quantities and the communication network interconnected to the analysis systems.

5.1.2. Block Diagrams of Primary Signals

The measurements required by SCADA and stored into database are performed by the remote stations RTU (Remote Terminal Units) and / or by signal capturing devices IED (Intelligent Electronic Devices). In the practice due to the unbalancing of loads, errors in measurements performed by SCADA and other factors which influence the electric system, it is verified that the amplitude values of quantities of interest (voltage and current) are different among the three phases of the transmission line.

This condition shows that the measured values bring contradiction levels among them right from the origin. So, in order to obtain reliable values in the signal treatment of the Paraconsistent Expert System -PES$_{PAL2v}$, the primary values receive an initial treatment of contradiction extraction.

Considering this condition, the extracting block of primary signals uses algorithms capable of extracting evidence degrees and of extracting contradiction effects, as shown on Figure 11.

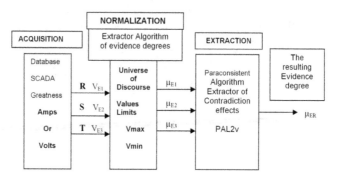

Figure 11. Block diagram which extracts primary evidence signals related to the sub-voltage and overcurrent on the measurement point.

The maximum and minimum values of the discourse universe of the evidence degree extracting algorithm are particular to each load connected to the corresponding breaker at the substation of the electric power transmission line.

The evidence degree extracting system receives the three values of voltage (or current) corresponding to the three phases of the line (RST) which are transformed in evidence degrees by a normalization defined by the interest interval or discourse universe. After this first process, the three resulting values pass through a contradiction effect extracting algorithm which outputs a unique resulting evidence degree.

5.2.3. Evidence Degree Extraction at an Operating Substation

PES$_{PAL2v}$'s project was carried out in order to perform analysis of overload risks through the applications of the algorithms of the PAL2v on monitoring essential points available at the operating substations of the sub-transmission electric system.

The buses that interconnect the several equipments installed at a substation such as transformers, electric keys and breakers of the sub-transmission system are points where voltage and current can be measured for each one of the loads interconnected by the breakers. In an operating substation a sub-voltage evidence degree and an overcurrent evidence degree are extracted from each breaker which activate loads at an operating substation.

Based on these values, the modules composed by the algorithms of the PAL2v verify the state of that point with respect to tension decreasing and excess of current intensity which together contributes to the increase of the overload risk at the point measured. After the extraction of the evidence degrees of sub-voltage and overcurrent from the load breaker, these two signals become input to a contradiction effect extracting algorithm outputting an evidence degree of overload risk at the measurement point.

Figure 12 shows the evidence degree of overload risk extracted from a measure point of the breaker (Load 16) of a typical substation of the sub-transmission system.

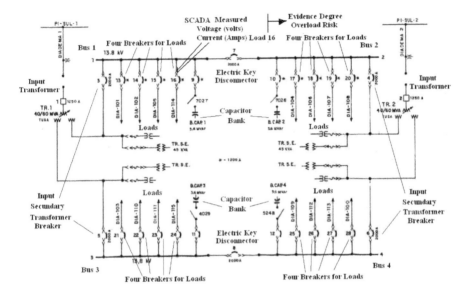

Figure 12. Overload Risk degree extraction from the breaker of load 16.

The operating substation seen on the figure 12 is composed by two buses shown horizontally where a total of 16 load control breakers are interconnected. Each breaker controls its correspondent load and are sources of extraction of evidence degrees for the paraconsistent analysis. In each one of the two buses there are 8 load breakers combined, however when a disconnector or electric key is open, it separates two buses with 4 breakers for each one.

Other breakers which control the feeding of the input transformers of the substation and the capacitor bank are also measurement points from where the evidence degrees of overload risk at the operating substation.

5.3. Extracting Module of Evidence Degrees from the Interconnection Buses

The extracting module of resulting risk evidence degree uses the contradiction effect extracting algorithm so it can receive n evidence degrees from several breakers which are interconnected to the same bus. That being, the PAL2v analysis can offer a unique representing value of the evidence degree of the bus.

Figure 13 shows a general diagram of capture of the resulting evidence degree of a bus at a typical operating substation shown in the previous picture which is composed by four breakers.

Through amplitude signals of the quantities received by the extraction modules of primary signals, and the signal which represents the state of the breaker key, each breaker has its evidence degree of risk which will be treated by the final module, resulting in a unique value of evidence degree of overload risk on the buses.

5.4. Paraconsistent Logical Model of Operating Substation

Using contradiction effect extracting blocks a paraconsistent logical model of operating substation can be created encompassing the evidence degree of overload risk of all possible points to be monitored.

In the typical operating substation the goal is to obtain the degree of overload risk generated by the four buses. So, from the models of the main devices installed in a typical substation, a paraconsistent modeling for the whole substation was designed.

First the overload risks at a typical substation were classified in four types:

1. Risks of overload on the buses (μ_{Ebus}).

2. Risks of load transfer (μ_{ETRANS}).

3. Risks of overload on the secondary windings (μ_{ESeC}).

4. Risks of overload on the primary windings (μ_{EPRIM}).

The resulting evidence degree of overload risk of the substation is then obtained by the paraconsistent analysis performed among these four values extracted from the model.

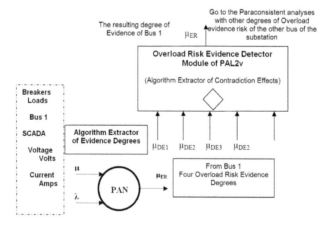

Figure 13. Diagram of capture of evidence degree of risks from a typical bus of a substation.

5.5. Computing the total resulting evidence degree of overload risk of the substation

After obtaining the evidence degree of overload risks the model for restoring control was developed. This model has as its goal to analyze and present optimized restoring procedures for the transmission of electric power after a contingency.

5.5.1. Restoring control actions

The restoring of an electric power system is a very complex procedure because it involves several activities which include: steps for previous study until steps of decision-making under intense emotional stress by the operators. The goal of the restoring control is to carry out the prompt restoration of the electric power system taking it to the condition of normal operation, where the load is attended and the operating limits are observed.

In the practice the reconnection procedure must be carried out by taking several precautions in order to suit all restrictions which take the new system state to a better level than that one which caused the disconnection. In order to start the actions of the restoring control, it is necessary to have full knowledge of the current situation of the electric system. The most important items which must be in the knowledge base of the model are:

a. Knowledge about the part of the electric power system which was disconnected.

b. Knowledge about the parts of the electric power system which were affected by the failure.

c. Knowledge of the source and cause of the disconnections, detecting what caused the disconnection with the best precision possible.

d. Knowledge of the existence of real conditions for the reconnection with the verification of the following situations:

If the source of the failure is a permanent failure that prevents from reconnecting.

If the disconnection happened from previously established emergency control actions. In this case the electric power system is known to have programmed "islands" which will make the restoring process easier.

If there are previously established control actions. This is a situation which happened when the emergency control takes the energy system to a condition known by previous studies.

5.6. Restoring Plans

The restoring plans have detailed actions by means of operation instructions which must be carried out by the operator in order to reconnect the electric power system. The strategy of these plans is based on the division of the procedures in steps in order to obtain a larger decentralization of the recomposition actions. In the development of the Paraconsistent Expert System - PES_{PAL2v}, the model of restoring control was designed to present actions based on the evidence degrees of overload risks obtained by the paraconsistent analysis on several points of the electric power system.

The reconnecting maps were done based on operational norms and restrictions of each substation and sequences of optimized restoring were established taking into consideration the values of risk degrees of overload before and after the contingency.

6. Implementation and Testing of the Paraconsistent Expert System - PES_{PAL2v}

The paraconsistent expert system PES_{PAL2v} was implemented to carry out analysis of the types of disconnection through information received by codes of the SCADA system and add this information to the restoring plan of the electric power system of an area being studied belonging to the AES-Eletropaulo Company which is an electric power concessionary company of Brazil.

The prototype was implemented on JAVA platform and tests with real values, which were extracted from the electric power lines of an area considered as a pilot and were stored in a history database, were carried out.

The project was in its first version developed so that the prototype PES_{PAL2v} performs off-line, however, with the information and data from events which represent real situations occurred in the area under study in the years 2007 and 2008. The pilot area, where the PES_{PAL2v} was tested with respect to its action analysis of overload risk and suggestions for the restoring, is composed by three OSs (operating substation), twelve TDSs (transformation and distribution station), twelve STCs (station of transformation to the consumer), three CBEs (capacitor bank station) and several aerial and underground lines.

The decision-making process of the SEP_{PAL2v} was designed through the acquisition of knowledge from the operators responsible for the electric operation in this area. When a contingency occurs, SEP_{PAL2v} receives the evidence degrees of overload risk through its paraconsistent logical model, performs a diagnosis and activates a flowchart with later available resources in performing the emergency maneuvers in the AES-Eletropaulo electric network considered as pilot.

6.1. Modeling and preparation of primary signals

Initially a large amount of data of the SCADA system related to that period was modeled to prepare the signals which are input to the prototype PES_{PAL2v}. The data stored in the SCADA system were modeled by creating two databases: the database of quantity values which will be called Database 1 and the database of alarms which will be called Database 2, as shown in Figure 14.

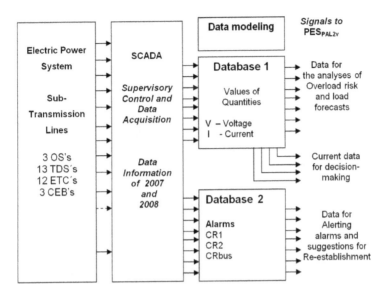

Figure 14. Initial data modeling diagram of the SCADA system.

Database 1 – Quantity values

After the data modeling of the SCADA system database 1 stores information about, besides those which identify the substation, breakers and other equipment and their measurements of amplitude of tensions and currents. The detections are so that in a time interval(Δt) PES_{PAL2v} is provided with measurements of intensities of currents in each load of the substation and the measurements of amplitude of tension on the buses in each stage of the load feeding, secondary an primary windings of the transformers.

Database 1 provides data for risk analysis that are the intensity of current of loads on the three buses (IA, IB, IC) and the amplitudes of tension of the bus on the three phases (VR, VS, VT). These values receive a paraconsistent logical treatment by PES_{LPA2v} such that the contradiction effects are decreased or totally excluded. Such contradictions are due to measurement mistakes inherent to the SCADA system. This treatment of the primary signals is performed by the special modules of capture and modeling.

Database 2 – Alarms

After the data modeling of the SCADA system database 2 stores information about, besides those which identify the substation, breakers (and other equipment), and types of classification of the alarms occurred in the events. The detections are so that in a time interval (Δt) PES_{PAL2v} is provided with the types of alarms that occurred in the installed component in the substation including the action of the relay keys (RC) with the types which classify the activated alarm: CR1, CR2 or Crbus.

Database 2 provides data for two purposes:

a) Data for detecting the topology - The data stored in the Database 2 represent the state (on or off) of the breakers and splitting keys. The signal of these states provides an overview of the topology of the substation which is transfered as evidence to the paraconsistent models of the breakers. The evidence degrees resulting from this analysis will influence the process of restoring suggestion generated by PES_{PAL2v}.

b) Data for detecting the occurrence type - The data stored in the Database 2 provide the alarm type and corresponding classification of the disruptions through several codes which are inserted on the restoring map of operating substation. The classification of the types of occurrences, together with the risk analysis signal PES_{PAL2v} the activation of the flowchart corresponding to the restoring map of the area affected by the contingency.

Figure 15 shows the signal flow where Database 1 and 2 are related with the modules of risk analysis, previewing and diagnosis.

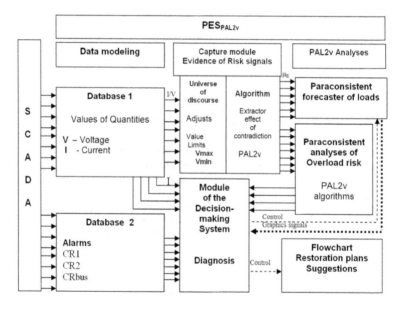

Figure 15. Signal flow between Database 1 and 2 and modules of paraconsistent analysis.

The decision-making module receives three types of signals: the values of intensities of the currents of the load (I) directly from Database 1; the values of the evidence degrees (μ_{ER}) from the paraconsistent analysis of overload risk; and signals from Database 2 related to the alarm types of occurrences. The analysis of these three signals results a diagnosis which activates a flowchart of restoring plan and the interaction with the user to find the best way to carry out the system restoring.

Based on the flowchart the steps for the restoring are following according to the diagnosis made based on the analysis which encompasses the classification of the alarm type, the values in engineering units of the measurements of currents and tensions, the risk evidence degrees obtained by the paraconsistent analysis carried out.

6.2. Verification of values on working screens

PES $_{PAL2v}$ has working screens where one can check the efficiency of the installed paraconsistent algorithms and the monitoring on each essential point of the substation.

Figure 16 below shows the values exposed on the screen of a typical substation (called "*Diadema*") used in the pilot system. On the screen one can see the evidence degrees of overload risks on all measurement points of the operating system being analyzed.

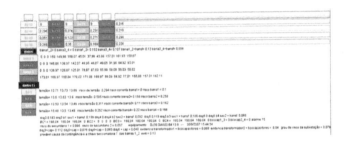

Figure 16. Working screen: significant values obtained from the Substation *Diadema*.

6.2.1. Denormalization Process

All the procedures for the analysis were carried out by the algorithms which were based on the PAL2v whose signal treatment considers normalized values, that is, values in the closed interval [0,1] of real numbers. In order to obtain previewing values in units of engineering, recovering the approximate values of current intensity and voltage, it is necessary to perform a denormalization process of the obtained values.

6.3. Tests and description of the application PES$_{PAL2v}$

The application of the Paraconsistent Expert System - PES$_{PAL2v}$ in this version can work in two modes according to the user: "Analysis" mode and "Training" mode. These two modes are described in what follows.

a) When the "Analysis" mode is selected, the Paraconsistent Expert System - PES$_{PAL2v}$ performs the analysis of overload risks, outputs the values of the risk degrees, current intensity and the breakers which are off. Next, the system interacts with the user and suggests optimized procedures for reconnecting. The suggestions are done in an interactive way through descriptions, visualization of the flowchart of the reconnecting maps and other restriction graphs.

b) When the mode "Training" is selected, the Paraconsistent Expert System - PES$_{PAL2v}$ will simulate the failure and step by step will present details about the procedures of the reconnecting flowchart. In order to begin the process the user has to inform the application the name of the substation he or she wants to simulate.

When this information is input, the application shows on the reserved space at the left of the screen the unifilar representation of the selected substation. The next step is the user's action which selects the breakers which will be simulated as "off" in order to configurate a type of failure occurrence.

When the simulation process is started the application, based on the breakers selected as "off" by the user, detects the type of alarm (CRs) which represents the disconnections and performs a search on the substation's database for the date that such failure occurred.

When the date of the occurrence is detected the application activates the networks of paraconsistent analysis obtaining the evidence degrees of overload risk and other specific information together with the first suggestions from the flowchart of the reconnecting map.

The interactive process is similar to the one presented in the "Analysis" mode: the suggestions and actions already determined by the flowchart will be step by step presented until the end of the optimized reconnecting. Doing so, the training is totally performed from real data of failure occurrences represented by values stored in the database.

Figure 17 shows a screen of an operating substation in its unifilar diagram with all available values obtained by the paraconsistent analysis. A menu, where restoring sequences of the electric power system after a contingency, is shown to the user.

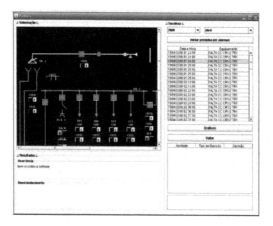

Figure 17. Analysis screen – Unifilar diagram and menu with information generated by PES$_{PAL2v}$.

7. Conclusion

In this work it was shown that the paraconsistent logics has a great capability of application in technological processes with the aim to solve complex problems. The Paraconsistent Expert System - PES_{PAL2v} was designed with an analyzing block of contingency which is capable of computing the risk degrees of outage by overloading of the electric power system. Moreover, given such occurrence, it is also capable of analyzing the conditions and of offering a list of sequences of optimized restoring for the operation.

Currently the expert system built with the PAL2v is being used to assist operation and training of operators at the operational substations of the electric power system of the AES-Eletropaulo – electric utility in Brazil. In the practice the paraconsistent expert system PES_{PAL2v} has shown to be an efficient tool, with which the user understands and accepts the reasoning methods used in the problem solving, since paraconsistent logics are more intuitive and has algorithms with simple structure. Generally speaking, it reached the following goals:

a. Assist the operator in the selection of the main control actions at the time of the restoring.

b. Outline and implement restoring plans based on the operational state of the electric system.

c. Show the restoring state in its optimized form.

Together to the above three main features, we can add three more:

d. Promote the operators' training.

e. Optimize the restoring process.

f. Detect "islands" – areas that due to disconnection remained isolated.

In operation, the PES_{PAL2v} has shown to be computational software where the modulation parameters are easy to adjust and the analyzing block of contingencies is adapted to provide resulting information in a satisfactory way. It was tested under several conditions using real values which were stored into a database for 12 months.

The sub-transmission system which was tested was composed of 12 substations where it was possible to modify and test several topological configurations. Under all tested conditions, PES_{PAL2v} showed good results and responded well to various situations in comparison to previous situations which were also stored into database.

The prototype application build in this first phase leave the necessary conditions fulfilled, so that the analysis process can be automatically started for the online implementation, topic which is for future projects. In this case, the alarms activated due to failures at the substations whose data were stored in the database, will start the application so that the analysis phases of the process are started in real-time. Under these conditions PES_{PAL2v} will with no doubt a very useful tool to the operation of the electric power system.

With this work it was shown that an expert system can be built with the algorithms of the paraconsistent logics and is capable of performing its fundamental task of analyzing contradicting information. Moreover, it is also capable to clearly show the user the reasoning methods it is using, so that the user can interact with the system with high confidence degree.

Author details

João Inácio Da Silva Filho[1*], Alexandre Shozo Onuki[1], Luís Fernando Pompeo Ferrara[1], Maurício Conceição Mário[1], José de Melo Camargo[2], Dorotéa Vilanova Garcia[1], Marcos Rosa dos Santos[2] and Alexandre Rocco[1]

*Address all correspondence to: inacio@unisanta.br

1 UNISANTA - Santa Cecília University, Brazil

2 AES - Eletropaulo Metropolitano Eletricidade de São Paulo S.A, Brazil

References

[1] CIGRE. (1993). Practical use of expert systems in planning and operation of power systems. TF 38.06.03, Électra, (146), 30-67.

[2] Paris, L., et al. (1984). Present Limits of Very Long Distance Transmission Systems. CIGRE. *Global Energy Network Institute*, Retrieved 2011-03-29.

[3] Donald, G. Fink., & Beaty, Wayne H. (2007). *Standard Handbook for Electrical Engineers (15th Edition)*, McGraw-Hill, 978-0-07144-146-9, section 18.5.

[4] Lynne, Kiesling. (2003). Rethink the Natural Monopoly Justification of Electricity Regulation. (August 18). Reason Foundation. Retrieved January 31, 2008.

[5] Da Silva Fiho, J. I., Lambert-Torres, G., & Abe, J. M. (2010). Uncertainty Treatment Using Paraconsistent Logic- Introducing Paraconsistent Artificial Neural Networks. *IOS Press*, 328, 211, Frontiers in Artificial Intelligence and Applications, Amsterdam, Netherlands.

[6] Da Silva Filho, J. I., Mário, M. C , Pereira, C. D. S. , Angari, A. C. , Ferrara, L. F. P. , Pitoli, O. Jr. , & Garcia, D. V. (2011). An Expert System Structured in Paraconsistent Annotated Logic for Analysis and Monitoring of the Level of Sea Water Pollutants. *Expert Systems for Human, Materials and Automation, Prof. PetricÄf Vizureanu (Ed.)*, 978-9-53307-334-7, InTech.

[7] Hughes, Thomas P. (1993). *Networks of Power: Electrification in Western Society, 1880-1930*, Baltimore: Johns Hopkins University Press, 119-122, 0-80184-614-5.

[8] Dieter Betz, Hans, Schumann, Ulrich, & Laroche, Pierre. (2009). *Lightning: Principles, Instruments and Applications*, Springer, 202-203, 978-1-40209-078-3, Retrieved on May 13.

[9] Da Silva Filho, J. I., Rocco, A., Mario, M. C., & Ferrara, L. F. P. (2006). Annotated Paraconsistent logic applied to an expert System Dedicated for supporting in an Electric Power Transmission Systems Re-Establishment. *IEEE PES- PSC 2006 Power System Conference and Exposition*, 212-220, 1-4244-0178-X, Atlanta, USA.

[10] Da Silva Filho, J. I., Rocco, A., Mario, M. C., & Ferrara, L. F. P. (2007). PES- Paraconsistent Expert System: A Computational Program for Support in Re-Establishment of The Electric Transmission Systems. *Proc. of LAPTEC 2007- VI Congress of Logic Applied to Technology*, 217, 978-8-59956-145-4, Santos, SP, Brazil, Nov.

[11] Da Silva Filho, J. I., Rocco, A., Onuki, A. S., Ferrara, L. F. P., & Camargo, J. M. (2007). Electric Power Systems Contingencies Analysis by Paraconsistent Logic Application. *Proc. of ISAP 2007- 14th International Conference on Intelligent System Applications to Power Systems*, 112-117, Kaohsiung, Taiwan, Nov.

[12] Da Silva Filho, J. I., & Rocco, A. (2008). Power systems outage possibilities analysis by Paraconsistent Logic. *Power and Energy Society General Meeting- Conversion and Delivery of Electrical Energy in the 21st Century*, IEEE, 978-1-4244-1905-0, 1932-5517, 1-6, Pittsburgh, PA.

[13] Torres, C. R., Abe, J. M., Lambert-Torres, G., & Da Silva Filho, J. I. (2011). Autonomous Mobile Robot Emmy III. *Mobile Robots- Current Trends*, Dr. Zoran Gacovski (Ed.), 978-9-53307-716-1, InTech.

[14] Da Costa, N. C. A., & Marconi, D. (1989). An overview of Paraconsistent logic in the 80's. *The Journal of Non-Classical Logic*, 6, 5-31.

[15] Da Costa, N. C. A. (1974). On the theory of inconsistent formal systems. *Notre Dame J. of Formal Logic*, 15, 497-510.

[16] Da Costa, N. C. A., Abe, J. M., & Subrahmanian, V. S. (1991). Remarks on annotated logic. *Zeitschrift f. math. Logik und Grundlagen d. Math*, 37, 561-570.

[17] Subrahmanian, V. S. (1987). On the semantics of quantitative logic programs. *Proc. 4th. IEEE Symposium on Logic Programming, Computer Society Press*, Washington D.C.

[18] Pereira, J. A. (2001). State Estimation Approach for Distribution Networks Considering Uncertainties and Switching. *PhD Thesis, FEUP*, Porto, July.

[19] Schulte, R., Sheble, G., Larsen, S., Wrubel, J., & Wollenberg, B. (1987). Artificial Intelligence Solutions to Power System Operating Problems. *IEEE Trans. On Power System*, Pwrs-2(4), Novembro de.

[20] Herbronn, B., Correa, J., Fandino, J., Hayashi, T., Kato, M., Krishen, D., Knight, U., Krost, G., Manolio, R., Matsumoto, K., Partanen, J., & Reichelt, D. (1993). A Survey of Expert Systems For Power System Restoration. *Electra* [No 150], Outubro de.

[21] Adibi, M., Clelland, P., Fink, L., Happ, H., Kafka, R., Raine, J., Scheurer, D., & Trefny, F. (1987). Power System Restoration- A Task Force Report. *IEEE Trans. On Power System*, Pwrs-2(4), Maio de.

Chapter 2

Intelligent Systems for the Detection of Internal Faults in Power Transmission Transformers

Ivan N. da Silva, Carlos G. Gonzales,
Rogério A. Flauzino, Paulo G. da Silva Junior,
Ricardo A. S. Fernandes, Erasmo S. Neto,
Danilo H. Spatti and José A. C. Ulson

Additional information is available at the end of the chapter

1. Introduction

This chapter presents an approach based on expert systems, which is intended to identify and to locate internal faults in power transformers, as well as to provide an accurate diagnosis (predictive, preventive and corrective), so that proper maintenance can be performed. In fact, the main difficulty in using conventional methods, based on analysis of acoustic emissions or dissolved gases, lies in how to relate the measured variables when there is an internal fault in a transformer. This kind of situation makes it difficult to design optimized systems, because it prevents the efficient location and identification of possible defects with sufficient rapidity. In addition, there are many cases where the equipment must be turned off for such tests to be carried out. Thus, this chapter proposes an architecture for an intelligent expert system for efficient fault detection in power transformers using different diagnosis tools, based on techniques of artificial neural networks and fuzzy inference systems. Based on acoustic emission signals and the concentration of gases present in insulating mineral oil and electrical measurements, intelligent expert systems are able to provide, as a final result, the identification, characterization and location of any electrical fault occurring in transformers.

With the changes occurring in the electricity sector, there is a special interest on the part of power transmission companies in improving and defining strategies for the maintenance of power transformers. However, when a fault occurs in a transformer, it is generally removed from the system and sent to a maintenance sector to be repaired. With this in mind, some feasibility studies have been conducted, aimed at supporting the electrical system in order to maintain the supply of energy, reducing operation costs and maintenance. Among these investigations, researches have been accomplished into the identification of internal faults in power transformers. In this case, the analysis of dissolved gases [1]-[5] and/or of acoustic emissions [6]-[10] can be highlighted. Within the context of economic viability, it is worth noting the increasing difficulty of removing an operating power transformer and placing it under maintenance. Thus, the above techniques, which evaluate parameters or quantities that indicate the current state of the transformer, have emerged as a more attractive alternative.

Although some papers deal with the development of tools for monitoring sensors [3], very few papers can be found on the efficient use of both sensor types (dissolved gases and acoustic emissions) in the same study. This is probably due to the fact that the cost associated with the acquisition of these sensors is very high. Another factor that should be highlighted is the growing use of intelligent tools for identifying and locating of internal faults [1-2, 5, 7].

The increasing use of intelligent tools is due to the fact that conventional techniques are not always able to achieve high accuracy rates of fault identification. In one of the most outstanding studies in the area [1], which makes a comparison between conventional and intelligent tools, the authors propose a method based on obtaining association rules that perform the best analysis of dissolved gases and satisfactorily ensure reliable identification of failures. The authors compared the proposed technique with other conventional methods (Rogers and Dornenburg) and intelligent techniques (Neural Networks, Support Vector Machines and k-Nearest Neighbors). A total of 1193 samples from dissolved gas sensors were acquired, which were divided into two sets of data in order to evaluate each technique used, i.e., one for training (1016 samples) and the other for validation (177 samples). After all training and validation processes had been conducted, the following accuracy rates were obtained: Artificial Neural Networks (62.43%), Support Vector Machines (82.10%), k-Nearest Neighbors (65.85 %), Rogers (27.19%), Dornenburg (46.89%) and Association Rules (91.53%). According to the results, it can be clearly seen that intelligent systems outperform conventional methods.

In addition to this paper, in [2], the authors make a more detailed analysis of gases. In this analysis, a total of 10 kinds of fault were considered, namely: partial discharge, thermal failures lower than 150°; thermal failures greater than 150° and lower than 200°; thermal failures greater than 200° and lower than 300°; cable overheating; current in the tank or iron core, overheating of contacts; low energy discharges, high energy discharges, continuous sparkling (a luminous phenomenon that results in the breakdown of the dielectric by discharge through the insulating oil), and partial discharge in solid insulation. It is worth mentioning that the method applied in this study was based on a fuzzy inference system, which was tested under controlled fault conditions. Other tests were also realized in Hungarian substation transmission transformers, where the method performed well against the uncontrolled failure scenarios.

However, studies [1] and [2] present a gap with regard to internal fault diagnosis for power transformers, because they only identify the type of failure and do not locate the partial discharges.

In order to provide a better fault diagnosis for power transformers, some studies have used acoustic emissions to locate faults due to partial discharges. Among these investigations, in [8], the authors propose a geometric analysis of the arrival times of acoustic emission signals in order to properly locate the sources of partial discharges. In the proposed methodology, they use both time measurements from sensors and pseudo-measurements, which provide greater precision in the tracking system of partial discharges.

In the context of these studies, this chapter aims to determine the necessary procedures for the development of a methodology based on information from sensors for both dissolved gases and acoustic emissions. The purpose of this methodology is achieve satisfactory results for identifying internal faults, and, in the case of faults due to partial discharges, to locate them accurately to help in the process of decision-making related to the maintenance of transmission transformers.

The tasks of identifying and locating internal faults in power transformers are extremely important, since they have a very high aggregate cost for purchase and for maintenance. Dissolved gas analysis and the analysis of partial discharges by means of acoustic emission sensors are essential for maintaining the equipment, and can bring many benefits, such as reducing the risk of unexpected failures, extending the useful life of a transformer, decreasing maintenance costs and reducing maintenance time (due to the precise location of the failure). Furthermore, with the processing of these data by means of intelligent expert systems, it becomes possible to provide answers to help in the decision-making process about the power transformer analyzed.

2. Internal Faults in Transformers

The diagnosis of the status and operating conditions of transformers is of fundamental importance in the reliable and economic operation of electric power systems. The aging and wear and tear of transformers determine the end of their useful life; thus, the occurrence of faults can affect the reliability or availability of the power transformer. Understanding the mechanisms of deterioration and having technically feasible and economically viable repair strategies enables us to correlate faults with the operating evolution of the equipment in service [11].

Many techniques have been proposed to ensure the integrity, reliability and functionality of power transformers, all of which seek trinomial low cost, efficiency and rapid diagnosis. Among several techniques available for detecting internal faults in power transformers, acoustic emission analysis can be highlighted because it is not invasive, allowing analysis to be conducted on the equipment during normal operation [12].

A power transformer can be affected by a variety of internal faults, such as partial discharge, electrical arcs, sparks, corona effects, and overheating. Of these, Partial Discharge (PD) can be highlighted, since it is directly related to the insulation conditions of a power transformer, which in turn trigger the occurrence of more severe faults. PD in high voltage systems occurs when the electric field and localized areas suffer significant changes which enable an electric current to appear [6].

According to [13], PD can be grouped into 8 classes:

- Point to Point discharges in insulating oil: these PDs are related to insulation defects between two adjacent turns in the winding of a transformer;

- Point to Point discharges in insulating oil with bubbles: this kind of fault is also caused by PD between two adjacent winding turns, but the condition of insulation degradation allows the formation of gas bubbles;

- Point to Plan in insulating oil: defects in the winding insulation system can cause PD between it and the grounded parts of the transformer tank;

- Surface Discharges between two electrodes: the most common kind of PD, occurring between two electrodes insulated with oil-paper called triple point, where the electrode surface is in contact with dielectric solids and liquids;

- Surface Discharges between an electrode and a multipoint electrode: the PD relating to these elements differ from the previous one with regard to the intensity distribution of the electric field. Both are insulated with oil-paper;

- Multiple Discharges on the plan: multiple damaged points in the winding insulation may cause PD between it and the grounded parts of the transformer tank;

- Multiple Discharges on the plan with gas bubbles: the PD in this case occurs at various damaged points in the winding insulation and the grounded parts of the transformer tank, but in the presence of gases dissolved in insulating oil;

- Discharges caused by particles: in this case, the insulating oil is contaminated with particles of cellulose fiber formed by the degradation process of the oil-paper insulation system, due to the aging of the power transformer. Such particles are in constant motion in the oil, causing PD;

3. Laboratory Aspects for Internal Fault Experiments in Power Transformers

It is important to specify equipment, methods and parameters, which vary according to the type of defect that is to be analyzed. In simple terms, the monitoring system can be better understood through Figure 1.

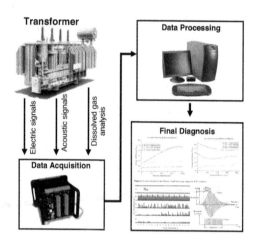

Figure 1. Laboratorial setup diagram.

The structures highlighted (inside the black boxes) are those that present the greatest challenges for configuration and parameterization, which are entirely dependent on the type of tests to be accomplished.

The most complete and detailed tests are (given their wide coverage of internal faults) more complex and expensive due to the various devices necessary used for the fault detection and location process, because more sensors and also data acquisition hardware are necessary.

3.1. Electrical measurements

Electrical parameters are also necessary for a correct characterization of internal transformer faults, especially when dealing with systems that require databases for normal operating conditions and with situations when a system has to be restored following a disturbance. This is the case of artificial neural networks, which require quantitative data for the learning process. It is necessary to measure voltages and three-phase primary and secondary currents, totaling 12 electrical parameters. The acquisition frequency in this case must not be high, because the purpose is to investigate the most predominant harmonic components in the electrical system.

3.2. Acoustic measurements

The acoustic signals are captured by acoustic emission sensors distributed evenly throughout the tank, which are externally connected to the power transformer. Such sensors have several characteristics that require a correct specification:

- Number of sensors per transformer: The number of sensors needed to detect internal faults in transformers varies according to the size of the equipment, amount of available channels and the type of fault to be detected. For the fault location task, for example, it takes a greater number of sensors, so that the entire volume of the transformer can be monitored. Thus, a total of 16 to 20 sensors is normally used [14];

- Pre-amplification: This item is extremely important because only the amplified acoustic signals are sent to the acquisition hardware, which removes extraneous noises;

- Operating frequency: This is strongly dependent on the type of fault to be monitored. Mechanical faults are associated with frequencies ranging from 20 kHz to 50 kHz, while electrical ones vary between 70 kHz and 200 kHz;

- Resonance frequency: This parameter defines the frequency where the signal gain is maximum. For maximum performance, it is necessary for the resonance frequency of the sensor to be tuned to the phenomenon to be monitored. The most common sensors have a resonance frequency of 150 kHz.

The experimental apparatus for supporting experiments aimed at testing computer systems developed for identifying and locating partial discharges in power transformers consists of a metal tank, in which all the devices responsible for the acquisition of acoustic and electrical signals are mounted. Figure 2 illustrates a tank specially prepared for this purpose.

Figure 2. Tank for experimental testing.

Figure 3 illustrates the attachment of an acoustic emission sensor mounted on the outside of the metal tank, whose signals are transmitted via cable to the acquisition system.

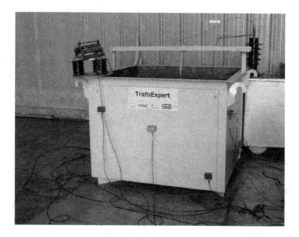

Figure 3. Acoustic emission sensor fixed to the outside of the tank.

Figure 4 illustrates a device made in order to produce partial discharges in the tank. The mechanism can also be moved within the tank, in all directions, by means of a rail and pulley system.

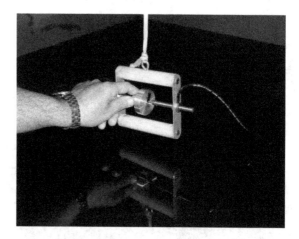

Figure 4. Device to produce partial discharges in the tank.

3.3. Measurements of dissolved gases

Measurement of dissolved gases in insulating oil can be acquired from chromatographic analysis of the oil, which is often performed in the laboratory. However, there are now some commercial devices that sense some gases dissolved in the oil. These devices can be used to monitor a power transformer in real time. It is worth mentioning that, through the analysis of dissolved gases, it is possible to obtain a first indication of a malfunction, which is usually related to electrical discharges and overheating.

Figure 5 shows the installation (in the tank) of the gas sensor, which is responsible for acquiring information on the quantities of gases dissolved in the insulating oil in order to relate them to internal defects.

Figure 5. Gas analysis sensor installed in the experimental tank.

3.4. Equipment for data acquisition

As seen above, the frequencies for electrical signals differ greatly from those found in acoustic signals, whose acquisition hardware can be divided into two according to technical and financial aspects:

- Hardware for electrical signals: for power quality purposes established in the Brazilian standard PRODIST [15], the 25th harmonic is the last one of interest. Thus, according to the Nyquist criterion, a minimal acquisition rate of 3 kHz is required. For electrical parameters it is also possible to use hardware with an A/D multiplexed converter, which reduces the cost of equipment;

- Hardware to acoustic signals: one of the factors that make this hardware expensive is the need to use an A/D converter for each channel. The sources of acoustic emissions also vary between 5 kHz and 500 kHz, where an acquisition frequency in MHz is necessary.

3.5. Computer for receiving and processing data

The computer is responsible for storing acoustic, electrical and dissolved gas data coming from the hardware acquisition. The hardware bus speed and the disk storage capacity must also take into account the amount of planned experiments, although a high performance disk is unnecessary, since a SCSI bus can be used.

3.6. Analysis and diagnosis

The implementation of this structure is very challenging, because it consists of a combination of techniques to efficiently identify and locate faults in power transformers. Among these techniques, those based on intelligent systems have efficiently increased the performance of processes involving the detection and location of faults [13].

4. Data Analysis from Acoustic Emission Signals

Altogether, we collected 72 oscillograph records of partial discharges. Each of these records depicts a time window of one second. In general, many occurrences of partial discharge are registered in these time slots.

In addition to this phenomenon, the data acquisition system also recorded mechanical waves that were used to evaluate the gauging of acoustic emission sensors. These waves are the result of the break, near the surface where the sensor is installed, of graphite with specifications given by the manufacturer of acoustic emission sensors. The graphs resulting from this test are highlighted in Figure 6.

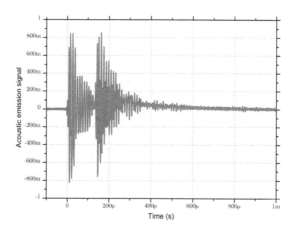

Figure 6. Acoustic emission signal resulting from the gauging process of sensors.

As shown in this figure, the signal is thus composed of two well-defined moments. The first of these relates to the instant when there was a mechanical disruption of graphite, while the second stage is the result of the impact of the pencil with the surface where the acoustic emission sensor is installed.

Figure 7 shows in more detail the first moment of the mechanical wave in Figure 6, while Figure 8 illustrates how the mechanical waves are related to the currents resulting from partial discharges.

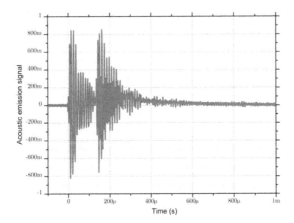

Figure 7. Details of the acoustic emission signal resulting from the gauging process of sensors.

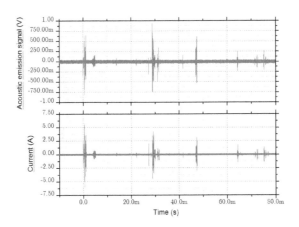

Figure 8. Relationship between partial discharge current and acoustic emission waves.

From Figure 8 we can see that each partial discharge results in a highly correlated mechanical wave. The graphs shown in Figure 9 highlight this relationship more clearly.

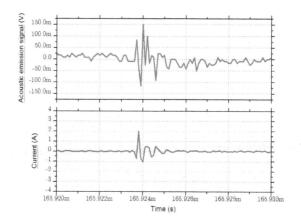

Figure 9. Detail of relationship between partial discharge current and acoustic emission.

Figure 10 illustrates the average frequency spectrum of an acoustic emission signal coming from a standard partial discharge. Through this frequency behavior, it can be seen that there is high signal energy at approximately 95 kHz and within the range between 160 kHz and 180 kHz. These values are of great importance in distinguishing partial discharge signals from other interferences.

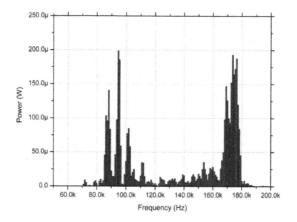

Figure 10. Average frequency spectrum of acoustic emission signal coming from a partial discharge.

In order to verify the behavior of the sensors for the tests, the voltage and current signals are processed in order to find the frequency response of these devices. In Figure 11 the amplitude versus frequency for the first calibration test has been recorded. The top of the graph highlights the energy and voltage signals sampled, and at the bottom there is the amplitude versus frequency. From the signal analysis it is then possible to observe a maximum response around 400 Hz and 100 kHz.

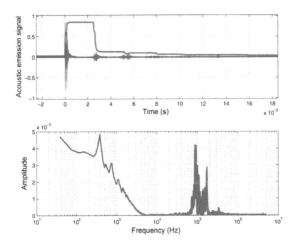

Figure 11. Frequency response of the acoustic emission signal.

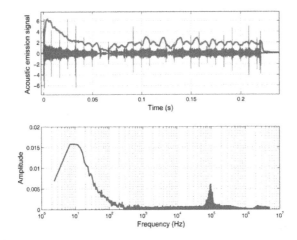

Figure 12. Detail of frequency response of the acoustic emission signal (segment 1).

In Figure 12, the signals were assigned in segments where the amplitude was more significant for detection purposes, which now represents the presence of different peak amplitudes at various frequencies.

The energy signal shows an envelope having important information, making clear the differences between the acoustic emission signal and the reflections that are also registered. In order to better evaluate these peaks, segments of interest were amplified and the frequency response was recalculated for this section, as reported in Figure 13.

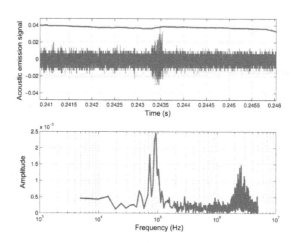

Figure 13. Detail of frequency response of the acoustic emission signal (segment 2).

In the segment highlighted in Figure 12, there is clearly a large concentration of low frequencies, with maximum amplitude at 10 Hz. In contrast, Figure 13 presents a large concentration at 100 kHz and another at approximately 2.5 MHz.

It is worth noting that, in the light of the two analyses, the signal with higher energy, recorded in the first segment, has an extremely low frequency wave. Thus, the propagation velocity tends to be higher due to the proximity to the spectrum of mechanical waves. However, for higher frequencies, typically observed in electromagnetic waves, there is a decrease of the signal energy, because this wave will suffer large attenuation when propagating through the insulating oil. Thus, the signal perceived by the acoustic emission sensor has already suffered severe degradation before being detected. This attenuation phenomenon is of great importance for the location process of partial discharges when installing more sensors in the experimental tank. In fact, since the speed of wave propagation in the insulating oil is known, it is then possible to estimate the location of the source of discharge.

The energy calculation is performed to obtain the full power of a signal. However, some signals are negative and therefore a quadratic sum of the sampled points must be calculated, as shown in the following equation:

$$E = \sum_{i=1}^{N} \sum_{j=1}^{M} sinal_{i,j}^{2} \qquad (1)$$

where N is the i-th window, and M represents the j-th point of the data window (consisting of 1101 points per window).

Thus, it may be noted that each data window corresponds to an acoustic emission signal measured by a given sensor. In this case, 8 sensors are used and, therefore, for each partial discharge we have 8 data windows. In addition, 10 samples for each partial discharge are still considered, which were obtained at different moments. Thus, the energy calculation for each of the 8 acoustic emission sensors is shown form Figures 14 to 21. Moreover, three different experiments were compared, where there was variation in the depth of the partial discharges in the oil tank used during the tests.

Experiment 1 represents a partial discharge located at 5 cm from the surface of the insulating oil, while experiments 2 and 3 are respectively located at 21.5 and 40 cm from the surface of the insulating oil.

Experiment 3 also had a small variation in the distance of the partial discharge from the front of the experimental tank, where it was moved 1 cm with respect to the original position of tests 1 and 2.

It is important to mention that this displacement is made in such away that the partial discharge of experiment 3 could be detected by sensors closer to the front wall of the tank, where it was expected that sensors 1 and 2 allocated on the wall would be more sensitive in experiment 3 rather than in experiments 1 and 2.

From Figures 14 and 15 it is possible to observe the energy response supplied by sensors 1 and 2 (for each of 10 samples), which represents the greatest contribution of experiment 1 in sensitizing them, while sensor 3 shows an energy response which makes it difficult to define which experiment caused the highest sensitization (Figure 16).

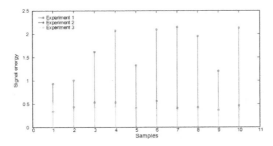

Figure 14. Energy response calculated for sensor 1 (mounted on the front wall - bottom right) during experiments 1, 2 and 3.

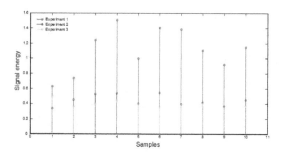

Figure 15. Energy response calculated for sensor 2 (mounted on the front wall - top left) during experiments 1, 2 and 3.

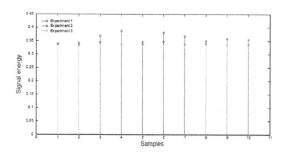

Figure 16. Energy response calculated for sensor 3 (mounted on the side wall-lower right corner) for experiments 1,2 and 3

The sensor 4 showed an energy response similar to that already shown for sensors 1 and 2 (Figure 17).

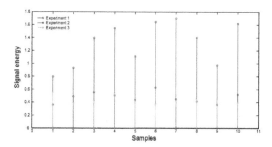

Figure 17. Calculated energy response for sensor 4 (mounted on the side wall - upper left) during experiments 1, 2 and 3

By means of the energy response supplied by sensor 5 (Figure 18) it can be seen that there is a certain emphasis on the response of experiment 1, but its energy levels are very close to those of experiments 2 and 3.

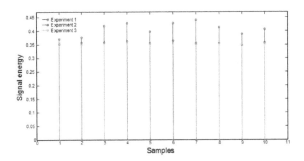

Figure 18. Energy response calculated for sensor 5 (mounted on the rear wall - bottom right) during experiments 1, 2 and 3.

The energy response of sensor 6 (Figure 19) in almost all samples presented responses similar to those obtained by sensors 1 and 2. However, in the first sample it can be seen that there are very similar levels of energy in the three experiments, although sensor 6 was a little more sensitive in experiment 3.

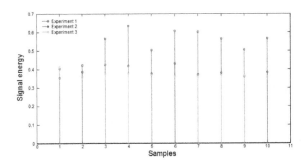

Figure 19. Energy response calculated for sensor 6 (mounted on the rear wall - top left) during experiments 1, 2 and 3

Sensor 7 presented the most complex energy response (Figure 20) because its response was unbiased for most samples. This is one factor that shows the complexity involved in the treatment of acoustic emission signals, making the application of intelligent systems very promising.

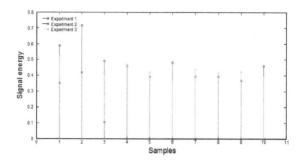

Figure 20. Energy response calculated for sensor 7 (mounted on the side wall - bottom left) during experiments 1, 2 and 3

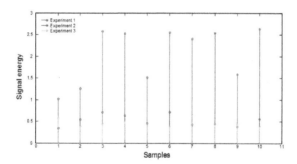

Figure 21. Energy response calculated for sensor 8 (mounted on the side wall - upper left) during experiments 1, 2 and 3.

Finally, sensor 8 presented an energy response (Figure 21) similar to that already obtained by other sensors, whose higher sensitization was caused by experiment 1.

5. Intelligent Systems

This section provides a theoretical foundation for fuzzy inference systems and artificial neural networks, as they are very prominent intelligent tools in the literature.

5.1. Fuzzy inference systems

Systems called fuzzy are built based on the theory of fuzzy sets and fuzzy logic, introduced by Zadeh in 1965, to represent knowledge from inaccurate and uncertain data. Fuzzy systems consist of a way to make a computational decision close to a human decision. Figure 22 shows a block diagram that expresses, in a simplified form, how a fuzzy system works.

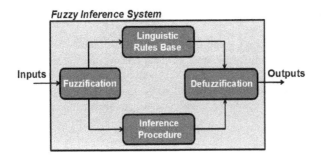

Figure 22. Diagram of a fuzzy inference system.

In the "Fuzzification" block, input values (in this case, information provided by the acoustic emission, gas concentration and electrical measurement sensors) are provided and conditioned, becoming fuzzy sets. Similarly, the "Defuzzification" block is responsible for transforming the outputs of the fuzzy system into non-fuzzy values (i.e., values which indicate the kind of internal fault and its location). The "Linguistic Rules Base" block has the function of storing the linguistic sentences and is fundamental to guarantee good system performance. The linguistic rules base and membership functions related to the inputs and outputs can be provided by experts or by automated methods, such as the ANFIS system (Adaptive Neural Fuzzy Inference System). On the other hand, the "Inference Procedure" block maps a system by using the linguistic rules. Thus, if rules are combined with input fuzzy sets acquired by the fuzzification interface, the system is then able to determine the behavior of the output variables of the system so that they can be defuzzified, generating the corresponding output to a given input value.

When using a fuzzy inference system, fuzzy rules and sets are adjusted and tuned by expert information. However, in some cases, because of the complexity and nonlinearity of the problem, it is necessary to use hybrid systems, such as ANFIS, where adjustments are performed in an automated manner according to the data set that represents the process. However, it is worth mentioning that, regardless of the setting, the whole fuzzy system has linguistic rules that can be represented as follows:

$$R_i : \mathbf{If} Input\ 1 \text{ is } x_1 \mathbf{and} Input\ 2 \text{ is } x_2$$
$$\mathbf{Then} Output \text{ is}$$
$$y_i = a \cdot x_1 + b \cdot x_2 + c$$

Another factor that should be noted is the inference procedure, in which a variety of methods can be used. Currently, the most commonly used methods are those of Takagi-Sugeno and Mamdani.

5.2. Artificial neural networks

Artificial neural networks are computational models inspired by the human brain, which can acquire and retain knowledge. Among the various neural network architectures, there is the architecture of multiple layers, called MLP (Multilayer Perceptron). This type of architecture is usually used for pattern recognition, functional approximation, identification and control tasks [16]. The structure of a neural network can be developed according to Fig. 3.

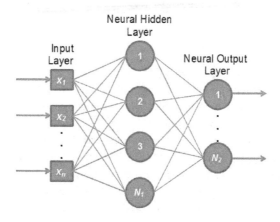

Figure 23. MLP neural network architecture.

As seen in Fig. 3, the neural network structure is basically composed of an input layer, hidden neural layers and an output neural layer. Also, between the layers, there is a set of weights, which are represented by a matrix of synaptic weights that will be adjusted during the training phase. It is further worth commenting that, for each of the neurons (hidden neural layers and output neural layer), it is necessary to implement activation functions in order to limit their output. In view of the basic configuration of the MLP neural network, other factors that should be explored are the training and validation stages.

During the training phase of MLP neural networks, some algorithms can be used. Currently, the backpropagation algorithm can be highlighted, which uses a descendent gradient calculation to reach the best adjustment of the synaptic weight matrix. In addition to the backpropagation algorithm, the Levenberg-Marquardt algorithm has been widely used because of its ability to accelerate the convergence process, due to the use of an approximation of Newton's method for non-linear systems [16].

On the other hand, the validation stage has the purpose of verifying the integrity of previously conducted training, so that the learning ability (generalization) of neural networks can be analyzed.

6. Intelligent Systems Used for the Identification and Location of Internal Faults in Power Transformers

As already mentioned in Section 1, a wide range of papers may be found in the literature, which are concerned with the identification and location of internal faults in transformers. However, there are very few papers which use intelligent systems applied to the same purpose, also taking into account experiments with acoustic emission sensors, electrical measurements and dissolved gases.

Among the most prominent papers found in the literature, we can highlight a few that use fuzzy inference systems and artificial neural networks for the analysis of dissolved gases [2, 17-19] and, for decision making, data from acoustic emission sensors [13].

As may be observed in papers [2, 17-18], which have fuzzy systems applied to the analysis of dissolved gases, the only notable difference lies in the fact that each one proposes different input variables to solve the problem and also different classes of faults. Thus, each paper has different settings of rules and of discourse universes for each input variable.

Therefore, a task of great importance is analyzing dissolved gases is the data preprocessing step, where the most relevant variables are obtained to characterize internal faults in power transformers.

As for those papers that analyze acoustic emission data, they typically employ conventional techniques [6-10]. However, the authors in [13] perform a series of experiments with partial discharges in insulating oil. However, these tests are not performed in order to apply the methodology to power transformers, but rather to identify partial discharges in any environment where oil is the insulator. Therefore, in order to identify the partial discharges, the authors use a MLP artificial neural network with backpropagation training, where the accuracy rates were above 97%.

Following the above context, it appears that the development of a method for identifying and locating internal faults in power transformers requires a number of steps, which are set out below:

- Allocation of sensors (acoustic emission and dissolved gases);

- Acquisition of data from sensors in accordance with the requirements commented upon in Section 3;

- Data preprocessing stage (definition of the most relevant variables and application of other necessary tools);

- Training or tuning of intelligent systems;

- Data validation (use of other data than those used in training/tuning stage);

- Performance analysis of the methodology in relation to other methodologies found in the literature.

It is worth mentioning that, out of the 6 steps mentioned above, most attention should be given to the allocation and acquisition of data, because bad data acquisition can affect the whole process of identifying and locating faults. It is also important to emphasize that the calculations made during the preprocessing of the signals was devised in order to extract the characteristics that best represent the positioning of the partial discharge in relation to the acoustic emission sensor. However, for this first stage of testing the expert system and the hardware used in the acquisition of the signals, we used the experimental tank.

In order to better represent the embedded software, a block diagram detailing the calculations to be performed by the software is set out below (Figure 24).

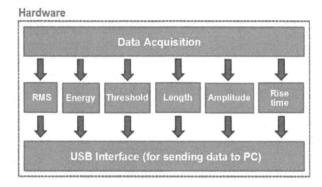

Figure 24. Overview of the embedded software.

As can be seen in Figure 24, it may be noted that the embedded software, after obtaining the acoustic signal, applies some computations in order to extract the characteristics that may represent the signal appropriately. Through these features, the expert system is able to distinguish these signals and to locate the source of partial discharges.

In this context, during the preprocessing step of the signs, the following calculations are performed: RMS, Energy, Length, Amplitude, Rise Time and Threshold. Finally, after obtaining the signal characteristics, they are sent to the computer through a USB (Universal Serial Bus).

Upon receipt of these data, the expert system is then responsible for providing information regarding the location of any partial discharge in the transformer. In order to better represent the overview of expert system, a block diagram is shown in Figure 25. In this figure, it may be noted that, after the received data concerning the characteristics commented upon previously, these are provided as input to the expert system.

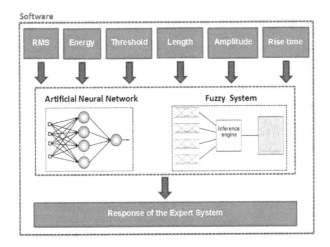

Figure 25. General diagram of the expert system.

In Figure 25 we can also observe that the expert system is composed of intelligent tools, such as artificial neural networks and fuzzy inference systems, which aim to locate partial discharges. Upon locating a partial discharge in transformer transmission, the operator may submit the equipment for maintenance (if necessary). Thus, the intelligent system here has the function of assisting the decision-making of the electric utility.

7. Conclusion

The tasks of identifying and locating internal faults in power transformers are extremely necessary, since this is one of the pieces of equipment that has the highest aggregated cost for both its purchase and maintenance.

Therefore, dissolved gas analysis and the analysis of partial discharges by means of acoustic emission sensors are essential for maintaining the equipment, which brings many benefits such as reducing the risk of unexpected failures and unscheduled downtime, extending transformer working life, reducing maintenance costs and minimizing maintenance time (due to failure location). Furthermore, processing this data by means of intelligent systems makes it possible to provide answers to help in decision-making about the analyzed power transformers.

Author details

Ivan N. da Silva[1*], Carlos G. Gonzales[2*], Rogério A. Flauzino[1], Paulo G. da Silva Junior[2], Ricardo A. S. Fernandes[1], Erasmo S. Neto[2], Danilo H. Spatti[1] and José A. C. Ulson[3]

*Address all correspondence to: insilva@sc.usp.br

1 University of São Paulo (USP), Brazil

2 São Paulo State Electric Power Transmission Company (CTEEP), Brazil

3 São Paulo State University (UNESP), Brazil

References

[1] Yang, Z., Tang, W. H., Shintemirov, A., & Wu, Q. H. (2009). Association Rule Mining-Based Dissolved Gas Analysis for Fault Diagnosis of Power Transformers. *IEEE Transactions on Systems*, Man, and Cybernetics- Part C: Applications and Reviews, 39(6), 597-610.

[2] Németh, B., Laboncz, S., & Kiss, I. (2009). Condition Monitoring of Power Transformers Using DGA and Fuzzy Logic. *Proceedings of the IEEE Electrical Insulation Conference EIC2009*, 31 May 2009-03June, Montreal, Canada.

[3] Snow, T., & Mc Larnon, M. (2010). The Implementation of Continuous Online Dissolved Gas Analysis (DGA) Monitoring for All Transmission and Distribution Substations. *In: Proceedings of the IEEE International Symposium on Electrical Insulation, ISEI*, 06-09June ,San Diego, USA.

[4] Szczepaniak, P. S., & Klosinski, M. D. G. (2010). DGA-based Diagnosis of Power Transformers- IEC Standard Versus k-Nearest Neighbors. *In: Proceedings of the IEEE International Conference on Computational Technologies in Electrical and Electronics Engineering, SIBIRCON*, 11-15 July, Listvyanka, Poland.

[5] Peng, Z., & Song, B. (2009). Research on Transformer Fault Diagnosis Expert System Based on DGA Database. *In: Proceedings of the 2nd International Conference on Information and Computing Science, ISIC*, 21-22 May, Manchester, UK.

[6] Mohammadi, E., Niroomand, M., Rezaeian, M., & Amini, Z. (2009). Partial Discharge Localization and Classification Using Acoustic Emission Analysis in Power Transformer. *In: Proceedings of the 31st International Telecommunications Energy Conference, INTELEC*, 18-22 October, Incheon, Korea.

[7] Veloso, G. F. C., Silva, L. E. B., Lambert-Torres, G., & Pinto, J. O. P. (2006). Localization of Partial Discharges in Transformers by the Analysis of the Acoustic Emission.

In: Proceedings of the IEEE International Symposium on Industrial Electronics, ISIE2009, 9-13July, Montreal, Canada.

[8] Markalous, S. M., Tenbohlen, S., & Feser, K. (2008). Detection and Location of Partial Discharges in Power Transformers Using Acoustic and Electromagnetic Signals. *IEEE Transactions on Dielectrics and Electrical Insulation*, 15(6), 1576-1583.

[9] Wang, X., Li, B., Roman, H. T., Russo, O. L., Chin, K., & Farmer, K. R. (2006). Acousto-optical PD Detection for Transformers. *IEEE Transactions on Power Delivery*, 21(3), 1068-1073.

[10] Núñez, A. (2006). Recent Case Studies in the Application of Acoustic Emission Technique in Power Transformers. *In: Proceedings of the IEEE/PES Transmission & Distribution Conference and Exposition: Latin America, TDC2006*, 15-18August Caracas, Venezuela.

[11] Flauzino, R. A., Silva, I. N., & Ulson, J. A. C. (2003). Neuro-Fuzzy Mapping of Dissolved Gases in Transformer Insulating Mineral Oil by Physico-chemical Tests (in Portuguese). *In: Proceedings of the VI Brazilian Symposium on Intelligent Automation, SBAI2003*, 14-17September Bauru, Brazil.

[12] Trindade, M. B., Martins, H. J. A., Cadilhe, A. F., & Moreira, J. A. (2005). On-Load Tap-Changer Diagnosis Based on Acoustic Emission Technique. *In: Proceedings of the XIV International Symposium on High Voltage Engineering, ISH2005*, 25-29 August, Beijing, China.

[13] Boczar, T., Borucki, S., Cichon, A., & Zmarzly, D. (2009). Application Possibilities of Artificial Neural Networks for Recognizing Partial Discharges Measured by the Acoustic Emission Method. *IEEE Transactions on Dielectrics and Electrical Insulation*, 16(1), 214-223.

[14] Trindade, M. B., Martins, H. J. A , & Menezes, R. C. (2008). Identification of Electrical and Mechanical Faults in Power Transformers and Reactors by Acoustic Emission (in Portuguese). *V International Workshop on Power Transformers, WORKSPOT2008*, 15-18April, Belem, Brazil.

[15] ANEEL. (2009). PRODIST- Establishes Procedures Related to Power Quality- PQ. Addressing Product Quality and Service Quality (in Portuguese), Brasilia: ANEEL.

[16] Silva, I. N., Spatti, D. H., & Flauzino, R. H. (2010). Artificial Neural Networks for Engineering and Applied Sciences- A Practical Course (In Portuguese). São Paulo: ArtLiber.

[17] Brescia, T., Bruno, S., La Scala, M., Lamonaca, S., Rotondo, G., & Stecchi, U. (2009). A Fuzzy-Logic Approach to Preventive Maintenance of Critical Power Transformers. *In: Proceedings of the 20th International Conference and Exhibition on Electricity Distribution, CIRED2009*, 8-11June, Prague, Czech Republic.

[18] Santos, L. T. B., Vellasco, M. B. R., & Tuscheit, R. (2009). Decision Support System for Diagnosis of Power Transformers. *In: Proceedings of the 15th International Confer-*

ence on Intelligent System Applications to Power Systems, ISAP2009, 08-12November, Curitiba, Brazil.

[19] Németh, B., Laboncz, S., & Kiss, I. (2010). Transformer Condition Analyzing Expert System Using Fuzzy Neural System. *In: Proceedings of the IEEE International Symposium on Electrical Insulation, ISEI2010*, 06-09June, San Diego, USA.

Research Applications

Chapter 3

Neural Networks and Decision Trees For Eye Diseases Diagnosis

L. G. Kabari and E. O. Nwachukwu

Additional information is available at the end of the chapter

1. Introduction

Clinical Decision Support Systems (CDSS) provide clinicians, staff, patients, and other individuals with knowledge and person-specific information, intelligently filtered and presented at appropriate times, to enhance health and health care [1]. Medical errors have already become the universal matter of international society. In 1999, IOM (American Institute of Medicine) published a report "To err is Human" [2], that indicated: First, the quantity of medical errors is incredible, the medical errors had already became the fifth lethal; Second, the most of medical errors occurred by the human factor which could be avoid via the computer system. Improving the quality of healthcare, reducing medical errors, and guarantying the safety of patients are the most serious duty of the hospital. The clinical guideline can enhance the security and quality of clinical diagnosis and treatment, its importance already obtained widespread approval [3]. In 1990, clinical practice guidelines were defined as "systematically developed statements to assist practitioner and patient decisions about appropriate health care for specific clinical circumstances" [4].

The clinical decision support system (CDSS) is any piece of software that takes as input information about a clinical situation and that produces as output inferences that can assist practitioners in their decision making and that would be judged as "intelligent" by the program's users [5].

Artificial intelligence has been successfully applied in medical diagnosis. They have been used for skin disease diagnosis, fetal delivery, metabolic synthesis as demonstrated in [6,7 and 8]. Artificial neural networks are artificial intelligence paradigms; they are machine learning tools which are loosely modelled after biological neural systems. They learn by training from past experience data and make generalization on unseen data. They have been applied as tools for modelling and problem solving in real world applications such as speech recognition, gesture recognition, financial prediction, and medical diagnostics [9, 10, 11 and 12]. Backpropagation employs gradient descent learning and is the most popular algorithm used for training neural networks. Neural networks were recently viewed as 'black boxes' as they could not explain how they arrived to a particular solution. Knowledge extraction is the process of extracting valuable information from trained neural networks in the form of 'if-then' rules as shown in [13 and 14]. The extracted rules describe the knowledge acquired by neural networks while learning from examples.

The human eye is the organ which gives us the sense of sight allowing us to learn more about the surrounding world than we do with any of the other four senses. We use our eyes in almost every activity we perform whether reading, working, watching television, writing a letter, driving a car and in countless other ways. Most people probably would agree that sight is the sense they value more than all the rest.

A recent survey of 1,000 adults shows that nearly half - 47% - worry more about losing their sight than about losing their memory and their ability to walk or hear. But almost 30% indicated that they don't get their eyes checked. Many Americans are unaware of the warning signs of eye diseases and conditions that could cause damage and blindness if not detected and treated soon enough.

In spite of the high prevalence of vision disorders in this country, so far, few victims receive professional eye care due to one of the following reasons;

• Specialist in eye diseases(ophthalmologist) are few and ophthalmology clinic are also few

• Lack of knowledge that early professional eye care is needed when symptoms are suspected.

• Inability to pay for the needed services.

Due to all of these, late detection of vision disorders and unnecessary loss of vision is encountered. But with a computer based system (expert system), over dependence on human expert can be minimized. This will go a long way to save time and furthermore early detection of eye disease can be adequately addressed. Cost for the services can also be reduced as a lot of unnecessary laboratory test may be avoided with the use of the proposed system.

This study classifies eye diseases using patient complaint, symptoms and physical eye examinations. The disease covered includes the following eye disease; Pink eye (conjunctivitis), Uveitis, Glaucoma, Cataract, Macular Degeneration, retinal detachment, Corneal ulcer, Color blindness, Far sightedness(hyperopia), Near sighteness(myopia), and Astigmatism.

We train artificial neural networks to classify eye diseases according to patient complain, symptoms and physical eye examination. We then use decision trees to extract knowledge from trained neural networks in order to understand the knowledge represented by the trained networks. Finally, we apply decision trees to build a tree structure for classification on the same sets of data sample we used to train neural networks earlier. In this way we combine neural networks and decision trees through training and knowledge extraction. The extracted knowledge from neural networks is transformed as rules which will help experts in understanding which combination of symptom, physical eye examination and patient's complain constituents have a major role in the eye problem. The rules contain information for sorting eye diseases according to their symptoms, physical condition and complain from the patient and knowledge acquired by neural networks from training on previous samples.

2. Application of Neural network in Clinical decision Support System

These days the Artificial Neural Networks(ANN) have been widely used as tools for solving many decisions modeling problems. The various capabilities and properties of ANN like Non-parametric, Non-linearity, Input-Output mapping, Adaptivity make it a better alternative for solving massively parallel distributive structure and complex task in comparison of statistical techniques, where rigid assumptions are made about the model. Artificial Neural Networks being non-parametric, makes no assumption about the distribution of data and thus capable of "letting the data speak for itself". As a result, they are natural choice for modeling complex medical problems where large database of relevant medical information are available [15].

In biomedicine, the assessment of vital functions of the body often requires noninvasive measurements, processing and analysis of physiological signals. Examples of physiological signals found in biomedicine include the electrical activity of the brain-the electroencephalogram (EEG), the electrical activity of the heart-the electrocardiogram (ECG), the electrical activity of the eye-i.e. PERG and EOG-respiratory signals, blood pressure and temperature signals [16].

Often, biomedical data are not well behaved. They vary from person to person, and are affected by factors such as medication, environmental conditions, age, weight, mental and physical state. Consequently, clinical expertise is often required for a proper analysis and interpretation of medical data. This has led to the integration of signal processing with intelligent techniques such as artificial neural networks (ANN), expert systems and fuzzy logic to improve performance [16].

ANN has been proposed as a reasoning tool to support clinical decision-making since 1959 [17]. Some problems encountered have led to significant developments in computer science, but it was only during the last decade of the last century that decision support systems have been routinely used in clinical practice on a significant scale [16].

The literature reports several applications of ANNs to the recognition of a particular pathology. For example, cancer diagnosis [18 and 19], automatic recognition of alertness and drowsiness from electroencephalography [20], predictions of coronary artery stenosis [21], analysis of Doppler shift signals [22 and 23], classify and predict the progression of thyroid-associated ophthalmopathy [24], diabetic retinopathy classification [25], saccade detection in EOG recordings [26] and PERG classification [22].

In this research we apply a hybrid of Neural Network and decision Tree to classify eye diseases according to patient complain, symptoms and physical eye examination. The aim is to help the ophthalmologist interpret the output of the examination systems easily and diagnose the problem accurately [27-29].

2.1. Artificial Neural Networks

Artificial Neural networks learn by training on past experience using an algorithm which modifies the interconnection weight links as directed by a learning objective for a particular application. A *neuron* is a single processing unit which computes the weighted sum of its inputs. The output of the network relies on cooperation of the individual neurons. The learnt knowledge is distributed over the trained networks weights. Neural networks are characterized into feedforward and recurrent neural networks. Neural networks are capable of performing tasks that include pattern classification, function approximation, prediction or forecasting, clustering or categorization, time series prediction, optimization, and control. Feedforward networks contain an input layer, one or many hidden layers and an output layer. Fig. 1 shows the architecture of a feedforward network. Equation (1) shows the dynamics of a feedforward network.

$$S^l{}_j = g_i\left(\sum_{i=1}^{m} S_i^{l-1} W^l_{ji} - \theta^l_j\right) \tag{1}$$

where $S^l{}_j$ is the output of the neuron j in layer$l$, $S_i^{l-1}$is the output of neuron j in layer l - 1 (containing m neurons) and W^l_{ji} the weight associated with that connection with j. $\theta^l_j$is the internal threshold/bias of the neuron and g_i is the sigmoidal discriminant function.

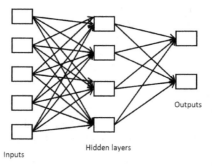

Figure 1. The architecture of the feedforward neural network with one hidden layer.

Backpropagation is the most widely applied learning algorithm for neural networks. It learns the weights for a multilayer network, given a network with a fixed set of weights and interconnections. Backpropagation employs gradient descent to minimize the squared error between the networks *output values* and *desired values* for those outputs. The goal of gradient descent learning is to minimize the sum of squared errors by propagating error signals backward through the network architecture upon the presentation of training samples from the training set. These error signals are used to calculate the *weight* updates which represent the knowledge learnt in the network. The performance of backpropagation can be improved by adding a momentum term and training multiple networks with the same data but different small random initializations prior to training. In gradient descent search for a solution, the network searches through a weight space of errors. A limitation of gradient descent is that it may get trapped in a local minimum easily. This may prove costly in terms for network training and generalization performance.

In the past, research has been done to improve the training performance of neural networks which has significance on its generalization. Symbolic or expert knowledge is inserted into neural networks prior to training for better training and generalization performance as demonstrated in [13]. The generalization ability of neural networks is an important measure of its performance as it indicates the accuracy of the trained network when presented with data not present in the training set. A poor choice of the network architecture i.e. the number of neurons in the hidden layer will result in poor generalization even with optimal values of its weights after training. Until recently neural networks were viewed as black boxes because they could not explain the knowledge learnt in the training process. The extraction of rules from neural networks shows how they arrived to a particular solution after training.

2.2. Knowledge Extraction from Neural Networks: Combining Neural Networks with Decision trees

In applications like credit approval and medical diagnosis, explaining the reasoning of the neural network is important. The major criticism against neural networks in such domains is that the decision making process of neural networks is difficult to understand. This is because the knowledge in the neural network is stored as real-valued parameters (weights and biases) of the network, the knowledge is encoded in distributed fashion and the mapping learnt by the network could be non-linear as well as non-monotonic. One may wonder why neural networks should be used when comprehensibility is an important issue. The reason is that predictive accuracy is also very important and neural networks have an appropriate inductive bias for many machine learning domains. The predictive accuracies obtained with neural networks are often significantly higher than those obtained with other learning paradigms, particularly decision trees.

Decision trees have been preferred when a good understanding of the decision process is essential such as in medical diagnosis. Decision tree algorithms execute fast, are able to handle a high number of records with a high number of fields with predictable response times, handle both symbolic and numerical data well and are better understood and can easily be translated into if-then-else rules.

The goal of knowledge extraction is to find the knowledge stored in the network's weights in symbolic form. One main concern is the fidelity of the extraction process, i.e. how accurately the extracted knowledge corresponds to the knowledge stored in the network. There are two main approaches for knowledge extraction from trained neural networks: (1) extraction of 'if-then' rules by clustering the activation values of hidden state neurons and (2) the application of machine learning methods such as decision trees on the observation of input-output mappings of the trained network when presented with data. We will use decision trees for the extraction of rules from trained neural networks. The extracted rules will explain the classification and categorization of different eye diseases according to symptoms.

In knowledge extraction using decision trees, the network is initially trained with the training data set. After successful training and testing, the network is presented with another data set which only contains inputs samples. Then the generalisation made by the network upon the presentation is noted with each corresponding input sample in this data set. In this way, we are able to obtain a data set with input-output mappings made by the trained network. The generalisation made by the output of the network is an indirect measure of the knowledge acquired by the network in the training process. Finally, the decision tree algorithm is applied to the input-output mappings to extract rules in the form of trees.

Decision trees are machine learning tools for building a tree structure from a training data-set of instances which can predict a classification given unseen instances. A decision tree learns by starting at the root node and selects the best attributes which splits the training data. The root node then grows unique child nodes using an entropy function to measure the information gained from the training data. This process continues until the tree structure is able to describe the given data set. Compared to neural networks, they can explain how they arrive to a particular solution. We will use decision trees to extract rules from the trained neural networks.

2.3. Decision Tree

A decision tree(DT) is a decision support tool that uses a tree-like graph or model of decisions and their possible consequences, including chance event outcomes, resource costs, and utility. It is one way to display an algorithm. Decision trees are commonly used in operations research, specifically in decision analysis, to help identify a strategy most likely to reach a goal. Another use of decision trees is as a descriptive means for calculating conditional probabilities.

Decision tree learning is a method commonly used in data mining. The goal is to create a model that predicts the value of a target variable based on several input variables. Each interior node corresponds to one of the input variables; there are edges to children for each of the possible values of that input variable. Each leaf represents a value of the target variable given the values of the input variables represented by the path from the root to the leaf.

A tree can be "learned" by splitting the source set into subsets based on an attribute value test. This process is repeated on each derived subset in a recursive manner called recursive partitioning. The recursion is completed when the subset at a node all has the same value of the target variable, or when splitting no longer adds value to the predictions. In data mining, trees can be described also as the combination of mathematical and computational techniques to aid the description, categorisation and generalisation of a given set of data. Data comes in records of the form:

$$(x, Y) = (x_1, x_2, x_3, \ldots, x_k, Y) \qquad (2)$$

The dependent variable, Y, is the target variable that we are trying to understand, classify or generalise. The vector x is composed of the input variables, x_1, x_2, x_3 etc., that are used for that task.

DT offers a structured way of decision making [29,30]. A DT is characterized by an ordered set of nodes. Each of the internal nodes is associated with a decision function of one or more features.. DT approach can generate *if -then* rules. Specific DT methods include Classification and Regression Trees (CART), Chi Square Automatic Interaction Detection (CHAID), ID3 and C4.5. C4.5 which is the extension of ID3[31,32] is very useful in this work. C4.5 Decision Tree is based on Information theory, that is it uses information theory to select features which give the greatest information gain or decrease of entropy [31]. Information gain is the informational value of creating a branch in a decision tree based on the given attribute using entropy theory.

2.4. Anatomy of the Eye

The eye is made up of numerous components. Figure 1 shows the anatomy of the eye.

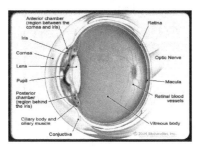

Figure 2. Anatomy of the eye (Source: http://www.medicinenet.com/eye_diseases_pictur_slideshow/article.htm#).

• Cornea: clear front window of the eye that transmits and focuses light into the eye

• Iris: colored part of the eye that helps regulate the amount of light that enters

• Pupil: dark aperture in the iris that allows light to go through into the back of the eye

• Lens: transparent structure inside the eye that focuses light rays onto the retina

• Retina: nerve layer that lines the back of the eye, senses light, undergoes complex chemical changes, and creates electrical impulses that travel through the optic nerve to the brain

• Macula: small central area in the retina that contains special light-sensitive cells and allows us to see fine details clearly

• Optic nerve: connects the eye to the brain and carries the electrical impulses formed by the retina to the visual cortex of the brain

• Vitreous: clear, jelly-like substance that fills the middle of the eye

2.5. Some eye disease conditions

Some eye disease conditions are shown in the figure3 to figure10 below:

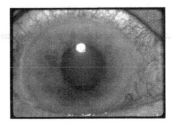

Figure 3. Glaucoma (Source: http://www.medicinenet.com/eye_diseases_pictur_slideshow/article.htm#).

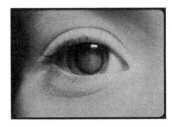

Figure 4. Cataract (Source: http://www.medicinenet.com/eye_diseases_pictur_slideshow/article.htm#).

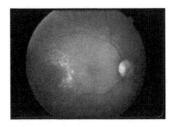

Figure 5. Macular degeneration (Source: http://www.medicinenet.com/eye_diseases_pictur_slideshow/article.htm#).

Figure 6. Conjunctivitis (Source: http://www.medicinenet.com/eye_diseases_pictur_slideshow/article.htm#).

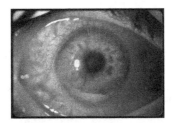

Figure 7. Uveitis (Source: http://www.medicinenet.com/eye_diseases_pictur_slideshow/article.htm#).

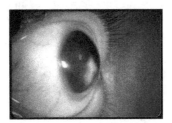

Figure 8. Keratoconus (Source: http://www.medicinenet.com/eye_diseases_pictur_slideshow/article.htm#).

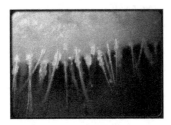

Figure 9. Blepharitis (Source: http://www.medicinenet.com/eye_diseases_pictur_slideshow/article.htm#).

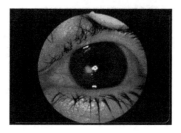

Figure 10. Corneal ulcer (Source: http://www.medicinenet.com/eye_diseases_pictur_slideshow/article.htm#).

3. Eye Diseases Diagnosis

We developed a clinical decision support system which bases its diagnosis on the patient complain, symptoms and physical eye examinations, and uses multilayer feedforward networks with a single hidden layer. Backpropagation algorithm is employed for training the networks in a supervised mode.

The eye diseases selected for diagnosis are as shown in table1. The designed neural network consists mainly of three layers: an input layer, a single hidden layer, and an output layer. The input layer has a total of 22 inputs plus the fixed bias input. These inputs consist of patient complaint, symptoms and physical eye examinations as may either be observed by the ophthalmologist or complained by the patients (i.e. X1, X2,..., X22). The output layer consists of 12 outputs indicating the diagnosed diseases (i.e. d1, d2,..., d13). Table1 shows the selected eye diseases for diagnosis and their symptom and signs as may be complained by patient or observed by the specialist while table2 shows the input variables for the system.

We ran 10 trial experiments with randomly selected 80% of the available data for training and the remaining 20% for testing the networks generalization performance. The learning rate of the network in gradient descent learning was 0.5. The network topology used was as follows: 22 neurons in the input layer for each symptom and signs for the eye disease, 9 neurons in the hidden layer and 13 neurons in the output layer representing each eye disease as shown in Fig. 2. We carried out some sample experiments on the number of hidden neurons to be used in the networks for this application. The results demonstrate that 9 neurons in the hidden layer were sufficient for the network to learn the training samples. The neural network was trained until one of the three following stopping criteria was satisfied:

1. On 100% of the training examples, the activation of every output unit was within 0.2 of the desired output, or

2. a network had been trained for 500 epochs, or

3. a network classified at least 98% of the training examples correctly, but had not improved it's ability to classify the training after ten epochs

S/N	Eye Disease	Signs (patient complain, symptoms and eye condition or cause)
1	Cataracts	Loss of visual acuity, loss of contrast sensitivity, contours shadows and color vision are less vivid, advanced age, bright light or antiglare sunglass may improve vision, poor night vision.
2	Glaucoma	Painless, decrease in peripheral field of view, halos around light, redness of eye, hereditary, aging may also cause it.
3	Macular Degeneration	Blurred vision, distorted images, missing letters in words, difficulty in reading, Trouble discerning colors, slow recovery of visual function after exposure to bright light, loss in contrast sensitivity, advanced age(66-74), Hereditary.
4	Pink eye(conjunctivitis)	Red or pink color eye, itching, blurred image, gritty feeling, irritation, watering of eye
5	Uveitis	Redness of eye, blurred vision, sensitivity to light(photophobia), dark floating sport in visual field, eye pain, blurred vision improves with blinking, discomfort after long period of concentrated use of eye(watching television, using computer or reading).
6	Retinal detachment	Experience of flashes of light and floater in visual view, feeling heaviness in the eye, central visual loss, blind spot in view.
7	Corneal ulcer	Redness of eye, pains of foreign bodies in the eye, pus/thick discharge from the eye, blurred vision, sensitivity to bright light, swollen eyelid, white or grey round spot on the cornea.
8	Keratoconus	Distorted vision, loss of vision focus, contact less could not improve vision
9	Blepharitis	Burning of foreign bodies sensation, itching, sensitivity to light, redness of eye, red and swollen eyelid, blurred vision, dry eye.
10	Color blindness	Problem discerning colors, hereditary, aging.
11	Farsightedness(hyperopia)	Blurred vision for close object, aging, contact lens may improves vision
12	Nearsightedness(myopia)	Blurred vision at distant, good vision for close object.
13	Astigmatism	Blurred vision, steamy appearing cornea, hereditary, may be corrected with contact lens

Table 1. Eye Diseases and their signs (patient complain, symptoms and eye condition).

Input variables	Variable Meaning
X1	Pains in the eye
X2	Redness or pink color of eye
X3	Bright light or antiglare sunglasses improves vision
X4	Poor night vision
X5	Family histories of the eye problem
X6	Decrease in peripheral field of view
X7	Age greater than 45 years
X8	Blurred vision
X9	Blurred vision improves with eye blinking
X10	Distorted vision
X11	Cloudy substance formed in front of eye lens
X12	Slow recovery of vision after exposure to bright light
X13	Irritation, itchy, scratchy or burning sensation of eye
X14	Discomfort after long concentration use of eye
X15	Trouble discerning colors
X16	Floaters in eye, flashes of light, halos around light
X17	Watering or discharge from eye
X18	Swelling of eye
X19	Steamy appearing cornea of eye
X20	Sensitivity to light (photophobia)
X21	Blurred vision for distant objects
X22	Blurred vision for close objects

Table 2. Input Variables and their Meaning.

The backpropagation algorithm with supervised learning was used, which means that we provide the algorithm with examples of the inputs and outputs we want the network to compute, and then the error (difference between actual and expected results) is calculated. The idea is to reduce this error, until the ANN learns the training data. The training begins with random weights, and the goal is to adjust them so that the error will be minimal. The activation function of the artificial neurons in ANNs implementing the backpropagation algorithm is given as follows[33]:

$$A_j(\bar{x}, \bar{w}) = \sum_{i=0}^{n} x_i \cdot w_{ji} \qquad (3)$$

$$O_j(\bar{x}, \bar{w}) = \frac{1}{\left[1 + e^{A(\bar{x}, \bar{w})}\right]} \qquad (4)$$

$$E_j(\bar{x}, \bar{w}, d) = \sum (O_j(\bar{x}, \bar{w}) - d_j)^2 \qquad (5)$$

$$\Delta w_{ji} = \eta \left(\frac{\partial E}{\partial w_{ji}}\right) \qquad (6)$$

Where: x_i are the inputs, w_{ji} are the weights, $O_j(x, w)$ are the actual outputs, d_j are the expected outputs and η - learning rate.

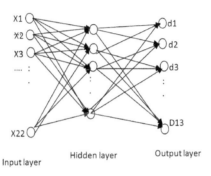

Figure 11. The neural network topology used for diagnosing the eye diseases which contain attribute information of 22 signs and symptoms. Each neuron in the input layer represents a particular sign or symptom. The neurons in the output layer represent the eye disease. Please note that all weight link interconnections are not shown in this diagram.

In this work, we used the C++ programming language in programming neural networks. Data mining and machine learning software tools such as 'Weka' can also be used for classification using neural networks.

4. Experimental Results

4.1. Data Set

The data set used for the training and testing of the system was collected from Linsolar Eye Clinic, Port Harcourt and Odadiki eye clinic, Port Harcourt all in Nigeria. The total data is 400 from which 320 samples (80%) are randomly chosen and used as training patterns and tested with 80 instances (20%) of the same data set. The data set consist of evenly distributed men and women. Samples also consider age randomly collected from 18 years to 70 years.

4.2. Rule Extraction from the ANN

As in Figure 12 it can be seen that both decision trees and neural networks can be easily converted into *IF THEN Rules* or we can simply convert neural networks into decision trees. In this work we use the networks architecture as shown in figure11 together with backpropagation algorithm with supervise learning.

Decision trees are machine learning tool for building a tree structure from a training dataset. A Decision tree learns by starting at the root node and select the best attributes which splits the training data [13]. Compared to neural networks they can explain how they arrive to a particular solution [34]. Hence, it usefulness in clinical decision support system as it may be use to support the expert in his delicate decision making or use as training tools for younger ophthalmologists. A typical decision tree extracted from the neural network in this work is shown in Figure 13.

To simplify complicated drawing the input variables that was shown in table1 may be combined to form conjunctions and negations which may also be used to generate the Decision Tree for some of the eye diseases as shown in Table 3.

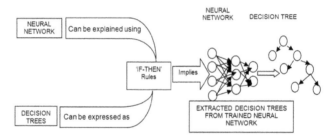

Figure 12. Extracting decision trees from neural networks.

Input variables	Variable Meaning	Input variables	Variable Meaning
NOT X1	No pains in eye	NOT X18	No swelling of eye
X1 and X2	Pains and redness of eye	X1 and X18	Pains and swelling of eye

Table 3. Some Additional variables for the Decision Tree.

The following rule sample sets are then obtained from the decision tree of Figure 13:

1. *IF* (X1 AND X2) and X18 and X20 and X8 and X11 and X17 *THEN* Cornel Ulcer

2. *IF* (X1 AND X2) and NOTX18 and X16 and X20 and X8 and X9 and X14 *THEN* Uveitis

3. *IF* (X1 AND X2) and NOTX18 and X16 and X6 and X5 and X7 *THEN* Glaucoma

4. *IF* NOTX1 and X8 and X10 and X5 and X12 *THEN* Muscular Degeneration

5. *IF* NOTX1 and X8 and X2 and X13 and X17 *THEN* Pink Eye

These rule sets are easily explain to means:

1. IF there is pains and redness of eye and swelling of eye and eye is sensitive to bright light and there is blurred vision and cloudy substance are formed in front of eye lens THEN the eye problem is *Cornel Ulcer*

2. IF there is pains and redness of eye and no swelling of eye and there are floater or flashes of light and eye is sensitive to bright light and there is blurred vision and the blurred vision improves with blinking of eye and there is discomfort after long concentrated use of eye THEN the eye problem is *Uveitis*

3. IF there is pains and redness of eye and no swelling of eye and there are floater or flashes of light and there are decrease in peripheral field of view and there is recorded family history of the eye problem and patient age is greater than 45 years THEN the eye problem is *Glaucoma*

4. IF there is no pains or redness of the eye and there is blurred and distorted vision and there is recorded history of the family history of the eye problem and slow recovery of vision after exposure to bright light THEN the eye problem is *Muscular Degeneration*

5. IF there is no pains of the eye and the eye is red and there is blurred vision and the eye is itchy or scratchy and there is watering discharge of the eye THEN the eye problem is *Pink Eye*.

Figure 13 shows an illustration of the extracted decision tree for some of the eye diseases.

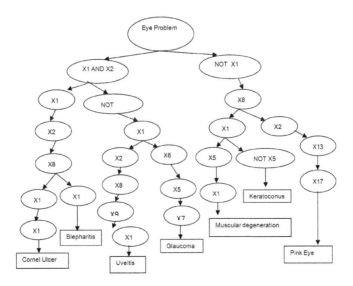

Figure 13. Decision tree for some the eye disease.

4.3. Performance Analysis of the system

To justify the performance of our diagnostic system, we conducted two analyses. The first is using a general performance scheme. Secondly, we carried out a number of tests at random using various physical eye examinations and patient's complain to see whether it agree with what it suppose to be.

4.3.1. Performance Benchmark

The proposed Neural Networks and Decision Tree for Eye Disease Diagnosis (NNDTEDD) architecture relies on a piece of software for easy eye disease diagnosis. The principles underlying diagnostic software are grounded in classical statistical decision theory. There are two sources that generate inputs to the diagnostic software: disease (H0) and no disease (H1). The goal of the diagnostic software is to classify each diagnostic as disease or no disease. Two types of errors can occur in this classification:

i.   Classification of disease as normal (false negative); and

ii.   Classification of a normal as disease (false positive).

We define:

Probability of detection $P_D = P_r$ (classify into H1 | H1 is true), or

Probability of false negative $= 1 - P_D$.

Probability of false positive $P_F = P_r$ (classify into H1 | H0 is true).

Let the numerical values for the no disease (N) and disease (C) follow exponential distributions with parameters λ_N and λ_C, $\lambda_N > \lambda_C$, respectively. Then we can write the probability of detection P_D and probability of false positive P_F as

$$P_D = \int_t^\infty \lambda_C e^{-(\lambda_C x)} dx = e^{-\lambda_C t} \tag{7}$$

$$P_D = \int_t^\infty \lambda_N e^{-(\lambda_N x)} dx = e^{-\lambda_N t} \tag{8}$$

Thus P_D can be expressed as a function of P_F as

$$P_D = P_F^r \tag{9}$$

Where $r = \lambda_F / \lambda_N$ is between 0 and 1.

Consequently, the quality profile of most detection software is characterized by a curve that relates its PD and PF, known as the receiver operating curve (ROC) [35]. ROC curve is a function that summarizes the possible performances of a detector. It visualizes the trade - off between false alarm rates and detection rates, thus facilitating the choice of a decision functions. Following the work done in [36], Figure 14 shows sample ROC curves for various values of r.

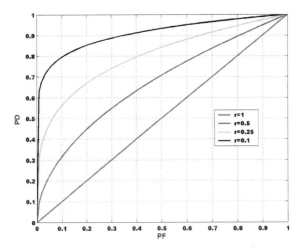

Figure 14. ROC curves for different r.

The smaller the value of r the steeper the ROC curve and hence the better the performance. The performance analysis of the NNDTEDD algorithms was carried out using MATLAB software package (MATLABR, 2009R) and the results compared with the collected data for cornel ulcer, uveitis, and glaucoma Figure 15, Figure 16 and Figure 17, respectively.

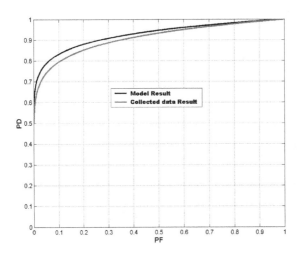

Figure 15. ROC curves for Cornel Ulcer diagnosis.

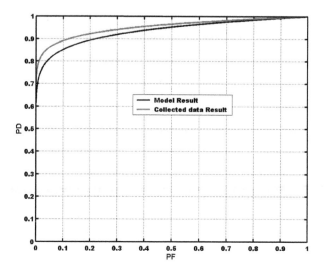

Figure 16. ROC curves for Uveities diagnosis.

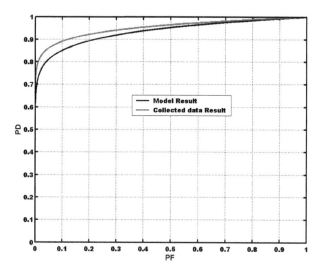

Figure 17. ROC curves for Glaucoma diagnosis.

4.3.2. Performance of the system using Random Tests

In testing the NNDTEDD, fifty different tests from the data sets for testing the system were carried out at random using various eye conditions and physical eye examinations combinations and the results compared with the expected result of the NNDTEDD. Where there was a match, success was recorded. In situations where there was no match failure were recorded. The total number of success = 46. Total number of failure = 4. Total number of test was 50.

$$Percentage\ succcess = \frac{46}{50} \times 100 = 92 \tag{10}$$

$$percentage\ failure = \frac{4}{50} \times 100 = 8\% \tag{11}$$

5. Conclusion

The research presented a framework for diagnosing eye diseases using Neural Networks and Decision Trees. This research extended common approaches of using a neural network or a decision tree alone in diagnosing eye diseases. We developed a hybrid model called Neural Networks Decision Trees Eye Disease Diagnosing System (NNDTEDDS). Neural networks have been successful in the diagnosis of eye diseases according to various symptoms and physical eye conditions. Decision trees have been useful in knowledge extraction from trained neural networks. They have been a means for knowledge discovery. We have obtained rules which explain the diagnosis of eye diseases according to various symptoms and physical eye conditions; these rules explain the knowledge acquired in neural networks by learning from previous samples of symptoms and physical eye conditions. The extracted rules can be used to explained how an eye disease is diagnosed hence removing the opacity in neural network alone. The extracted rules can also be used to train younger ophthalmologists. The proposed system was able to achieve a high level of success using the hybrid model of neural networks and decision tree technique. A success rate of 92% was achieved. This infers that combination of neural networks and decision tree technique is an effective and efficient method for implementing diagnostic problems.

6. Recommendations

This work is recommended to medical experts (ophthalmologists) as an aid in the decision making process and confirmation of suspected cases. Also, a non expert will still find the work useful in areas where prompt and swift actions are required for the diagnosis of a given eye disease covered in the system. Medical practitioners who operate in areas where there are no specialist (ophthalmologist) can also rely on the system for assistance.

Acknowledgements

We thank Dr. Monday Nkadam of the University of Port Harcourt Teaching Hospital for checking and correcting some of the symptoms that were used in the system. The authors also thank the staff of Linsolar and Odadiki eye clinic, Port Harcourt for using their data in training and testing the system.

Author details

L. G. Kabari[1*] and E. O. Nwachukwu[2]

*Address all correspondence to: ledisigiokkabari@yahoo.com

1 Department of Computer Science, Rivers State Polytechnic, Bori, Nigeria

2 Department of Computer Science, University of Port Harcourt, Port Harcourt, Nigeria

References

[1] Osheroff, J. A., Teich, J. M., & Middleton, B. F. (2006). A Roadmap for National Action on Clinical Decision Support. American Medical Informatics Association; 2006 June 13. Available at: http://www.amia.org/inside/initiatives/cds/. Accessed March 20, 2009.

[2] Kohn, L. T., Corrigan, J. M., & Donaldson, M. S. (2000). To err is human: building a safer health system. Washington, D.C.: National Academy Press.

[3] Miller, M., & Kearney, N. (2004). Guidelines For Clinical Practice: Development, Dissemination and Implementation. *International Journal of Nursing Studies*, 41(7), 813.

[4] Field, M. J., & Lohr, K. N. (2005). Clinical Practice Guidelines: Direction for a New Program. Institute of Medicine, Committee on Clinical Practice Guidelines. Washington, DC. National Academy Press.

[5] Musen, M. A. (1997). Modelling of Decision Support. *Handbook of medical informatics*, J. H. V. Bemmel and M. A. Musen, Eds. Houten: Bohn Stafleu Van Loghum.

[6] Bakpo, F. S., & Kabari, L. G. (2011). Diagnosing Skin Diseases using an Artificial Neural Network. *Artificial Neural Networks- Methodological Advances and Biomedical Applications, kenji suzuki (ed.)*, 978-9-53307-243-2, intech, Available from: http://www.intechopen.com/articles/show/title/diagnosing-skin-diseases-using-an-artificial-neural-network.

[7] Janghel, R. R., Shukla, A., Tiwari, R., & Tiwari, P. (2009). Clinical Decision support system for fetal Delivery using Ar tificial Neural Network. *2009 International Conference on New Trends in Information and Service Science.*

[8] Zhou, Q. (2009). A Clinical Decision Support System for Metabolism Synthesis. *2009 International Conference on Computational Intelligence and Natural Computing.*

[9] Robinson, A. J. (1994). An Application of Recurrent Nets to Phone Probability Estimation. *IEEE Transactions on Neural Networks,* 5(2), 298-305.

[10] Giles, C. L., Lawrence, S., & Tsoi, A. C. (1997). Rule Iinference for Financial Prediction using Recurrent Neural Networks. *Proceedings of the IEEE/IAFE Computational Intelligence for Financial Engineering,* New York City, USA, 253-259.

[11] Marakami, K., & Taguchi, H. (1991). Gesture Recognition Using Recurrent Neural Networks. *Proceedings of the SIGCHI conference on Human factors in computing systems: Reaching through Technology, Louisiana, USA,* 237-242.

[12] Omlin, C. W., & Snyders, S. (2003). Inductive Bias Strength In Knowledge-Based Neural Networks: Application to Magnetic Resonance Spectroscopy of Breast Tissues. *Artificial Intelligence in Medicine,* 28(2).

[13] Chandra, R., & Omlin, C. W. (2007). Knowledge Discovery Using Artificial Neural Networks For A Conservation Biology Domain. *Proceedings of the 2007 International Conference on Data Mining, Las Vegas, USA,* In Press.

[14] Fu, L. (1994). Rule Generation from Neural Networks, IEEE Transactions on Systems. *Man and Cybernetics,* 24(8), 1114-1124.

[15] Niti, G., Anil, D., & Navin, R. (2007). Decision Support System For Heart Disease Diagnosis Using Neural Networks. *Delhi Business Review,* 8(1), January- June 2007.

[16] Lisboa, P. J. G., Ifeachor, E. C., & Szczepaniak, P. S. (2000). Artificial Neural Networks In Biomedicine. London: Springer-Verlag.

[17] Ledley, R. S., & Lusted, L. B. (1959). Reasoning Foundations of Medical Diagnosis. *Science,* 130, 9-21.

[18] Abbass, H. A. (2002). An Evolutionary Artificial Neural Networks Approach for Breast Cancer Diagnosis. *Artificial Intelligence in Medicine,* 25, 265-281.

[19] Zhou, Z. H., Jiang, Y., Yang, Y. B., & Chen, S. F. (2002). Lung Cancer Cell Identification Based On Artificial Neural Network Ensembles. *Artificial Intelligence in Medicine,* 24, 25-3.

[20] Vuckovic, A., Radivojevic, V., Chen, A. C. N., & Popovic, D. (2002). Automatic Recognition of Alertness and Drowsiness from EEG by an Artificial Neural Network. *Medical Engineering & Physics,* 24, 349-360.

[21] Mobley, B. A., Schechter, E., Moore, W. E., McKee, P. A., & Eichner, J. E. (2000). Predictions of Coronary Artery Stenosis by ANN. *Artificial Intelligence in Medicine,* 18, 187-203.

[22] Kara, S., Gu°ven, A., Okandan, M., & Dirgenali, F. Utilization of Artificial Neural Networks and Autoregressive Modeling in Diagnosing Mitral Valve Stenosis. (in press), *Computers in Biology and Medicine*.

[23] Wright, I. A., Gough, N. A. J., Rakebrandt, F., Wahab, M., & Woodcock, J. P. (1997). Neural Network Analysis of Doppler Ultrasound Blood Flow Signals: A pilot study. *Ultrasound in Medicine and Biology*, 23(5), 683-690.

[24] Salvi, M., Dazzi, D., Pelistri, I., Neri, F., & Wall, J. R. (2002). Classification and Prediction of the Progression of Thyroid-Associated Ophthalmopathy By An Artificial Neural Network. *Ophthalmology*, 109(9), 1703-8.

[25] Nguyen, H. T., Butler, M., Roychoudhry, A., Shannon, A. G., Flack, J., & Mitchell, P. (1996). Classification of Diabetic Retinopathy Using Neural Networks. *18th Annual International Conference of the IEEE Engineering In Medicine And Biology Society, Amsterdam*, 1548-1549.

[26] Tigges, P., Kathmann, N., & Engel, R. R. (1997). Identification of Input Variables for Feature Based Artificial Neural Networks-Saccade Detection in EOG Recordings. *International Journal of Medical Informatics*, 45, 175-184.

[27] Chan, B. C. B., Chan, F. H. Y., Lam, F. K., Lui, P. W., & Poon, P. W. F. (1997). Fast Detection of Venous Air Embolism in Doppler Heart Sound Using the Wavelet Transform. *IEEE Transactions on Biomedical Engineering*, 44(4), 237-245.

[28] Tu°rkog˘lu, I., Arslan, A., & I˙lkay, E. (2002). An Expert System for Diagnosis of the Heart Valve Diseases. *Expert Systems with Applications*, 23, 229-236.

[29] Schmitz, G. P. J., Aldrich, C., & Gouws, F. S. (1999). ANN-DT An Algorithm for Extraction of Decision Trees from artificial Neural Networks. *IEEE Transactions on Neural Networks*.

[30] Sethi, I. K. (1992). Layered Neural Net Design Through Decision Trees. *Proceedings of the IEEE*.

[31] Quilan, J. (1993). C4.5: Programs for Machine Learning. Morgan Kaufmann, San Mateo, CA.

[32] Fabian, H. P., Chan, K. S., Ho, K. Y., & Leong, S. K. (2004). A Study on Decision Tree. *2nd Engineering & Technology Student's Congress. Kota Kinabalu.: SKTM*.

[33] Haykin, S. (1999). Neural Networks: a Comprehensive Foundation. Second Edition. Prentice Hall., 842.

[34] Yedjour, D., Yedjour, H., & Benyettou, A. (2011). Explaining Results Of Artificial Neural Networks. *Journel Of Applied Scinces*, 2(3).

[35] Trees, H. V. (2001). Detection, Estimation and Modulation Theory- Part I. John Wiley, New York.

[36] Huseyin, C., & Srinivasan, R. (2004). Configuration of Detection Software: a Comparison of Decision and Game Theory Approaches. *Decision Analysis*, 1(3), 131-148.

Chapter 4

Neuro-Knowledge Model Based on a PID Controller to Automatic Steering of Ships

José Luis Calvo Rolle and Héctor Quintián Pardo

Additional information is available at the end of the chapter

1. Introduction

In the area of control engineering, it is necessary to work in a continued form in obtaining new methods of regulation to remedy deficiencies that already exist, or to find better alternatives to which they were used previously, for example [1, 2]. This dizzying demand of applications in control, is due to the wide range of possibilities developed until this moment.

Despite this upward pace of discovery of different possibilities, it has been impossible up to now to derail relatively popular techniques, such as the "traditional" PID control. Since the discovery of such regulators by Nicholas Minorsky in the area of automatic steering of ships [3, 4] since 1922 by now, many studies on this regulator are made. It should be noted that there are numerous usual control techniques for the process in any field, where innovations were introduced, for example thanks to the inclusion of artificial intelligence in this area, [5]. Despite this, the vast majority of its implementation uses PID controllers, raising the utilization rate up to 90% [6], according to different authors. Their use continues to be very high for different reasons such as ruggedness, reliability, simplicity, error tolerance, etc.

In the conventional PID control [6, 7] there are many contributions made by scholars as a result of investigations conducted on the subject,. There are many expressions among them, for obtaining the parameters that define this control, achieved by different routes and operating conditions specific for the plant that tries to control.

Emphasize that formulas developed to extract terms, which are sometimes empirical, always go out to optimise a particular specification. On the other side often happens that when one of the parameters is improved, other gets worse. We need to indicate that the parameters obtained applying the formulas of different authors are a starting point for setting the regulator. Usually it is necessary to proceed afterwards to a finer tuned test-error.

The vast majority of real systems are not linear. This occurs notably in the steering of ships, from which long time ago emerging models as the first or second order Nomoto model [8], the Norbbin model [9] or the Bech model [10]. Nowadays this feature remains a cause for study [11, 12] according to their working point, certain specifications will be required to be equal in all areas of operation. Thus different values of the regulator parameters will be needed in each of these areas. Having this in mind, self and adaptive PID regulators [5, 13–15] are a good solution to reduce this problem. Though it should be noted that its implementation is quite difficult, expensive and closely linked to the type of process which purports to regulate, being sometimes difficult to establish a general theory in this type of PID controllers.

To alleviate these difficulties it can be applied the well-known Gain Scheduling method, which is easier to implement, and with which are obtained highly satisfactory results. The concept of Gain Scheduling arises at the beginning of the 90t's [16], and it is considered as part of the family of adaptive controllers [13]. The principle of this methodology is to divide a non-linear system in several regions in which its behaviour is linear. Thus we obtain parameters of the controller that allow having some similar specifications around the operating range of the plant.

To implement the Gain Scheduling at first it is necessary to choose the significant variables of the system according to which it is going to define the working point. Then it is necessary to choose operating points along the entire range of operation of the plant. There is no systematic procedure for these tasks. Often at first step are taken those variables that can be measured easily. The second step is more complicated because of the points that have to be selected. The system can be stable at them for the parameters of the controller deducted, but it does not have to be stable between the selected points. This problem has no simple solution, and when it exists, there is usually particularized, that is why a subject has been studied by researchers, see for example [17–19].

A way to solve the problem is using artificial neural networks, which is a known side of the Artificial Intelligence that is in general difficult and uses other techniques. There are some similar cases where their work is to be resolved by this technique [20, 21], as well as other techniques of artificial intelligence [22, 23]. As it will be shown throughout this document, the use of the proposed method may be feasible in many cases.

This document is structured starting with a brief introduction of the topology of PID controller, which it is used to show an explanation of the method proposed above. Then is exposed a description of its application to the model of a ship, which is a non-linear system, to carry out the steering control with the proposed methodology. That is applicable in different steering models of existing ships regardless of the complexity. It ends with the validation of the method, making simulations under different conditions.

2. The PID controler

There are multiple representation forms of PID controller, but perhaps the most widespread and studied is the one given by the equation 1

$$u(t) = K\left[e(t) + \frac{1}{T_i}\int e(t)dt + T_d\frac{de(t)}{dt}\right] \tag{1}$$

where u is the control variable and y is the control error given by $e = YSP - y$ (difference between the reference specified by the input and the output measured in the process). Thus, the control variable is a sum of three terms: the term P, which is proportional to the error, the term I, which is proportional to the integral error, and the term D, which is proportional to the derivative of error. The controller parameters are: the proportional gain K, the integral time Ti and the derivative time $T - d$.

There are multiple ways for the representation of a PID controller, but to implement the PID controller used and defined in the formula above, and more commonly known as the standard format [6, 7], shown in representation bloc, it is shown in Figure 1.

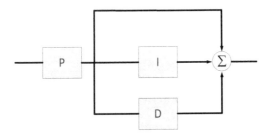

Figure 1. PID controller in standard format

There are infinite processes that exist in industries whose normal function is not adequate for certain applications. The problem is often solved by using this controller, by which the system is going to obtain certain specifications in the process control leading them to optimal settings for the certain process. The adjustment of this controller is carried out by varying the proportional gain, and the integral and derivative times commented in its different forms.

3. Adjustament methods of parameters controller with gain scheduling

On many occasions this method is known as the process dynamic changes with the process operating conditions. One reason for the changes in the dynamic can be caused, for example, by the well-known nonlinearities of processes. Then it will be possible to modify the control parameters, monitoring their operation conditions and establishing rules. The methodology will consists of first application of Gain Scheduling, analyzing the behaviour of the plant in question at different points of work and establishing rules to program gains in the controller, so that it will be possible to obtain certain specifications which remain, in the possible extent, constant throughout the whole range of operation of the process. This idea can be schematically represented as shown in Figure 2.

The Gain Scheduling method can be considered as a non-linear feedback of a special type; it has a linear controller whose parameters are modified depending on the operation conditions, with some rules extracted and previously programmed. The idea is simple, but its implementation is not easy to carry out, except in computer controlled systems. As it is shown in Figure 3, operating conditions that indicate the working point that is the process, with the specific rules learned, program in the controller, the parameters selected.

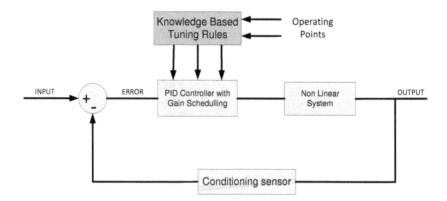

Figure 2. Gain Scheduling control schematic

4. Implementation with neural networks instead of the knowledge base tuning rules

The replacement of the rules that define the gains of PID controller based on the working point system is raised by a neural network whose inputs are the operation conditions of the plant, and as outputs it will have the parameters of PID controller (K, Ti and Td). The neural network used is a network type MLP (Multi Layer Perceptron) with a similar topology to the one shown in Figure 3.

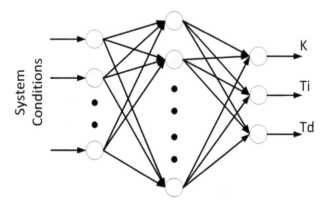

Figure 3. Neuronal network structure

It is possible to raise several simple neural networks for each parameter instead of one, in which all of them have as input, the system conditions but the output is the controller parameter in each case.

5. Nomoto model of ship-steering process

To analyze a ship's dynamics as a Nomoto model it is convenient to define a coordinate system as indicate in figure 4.

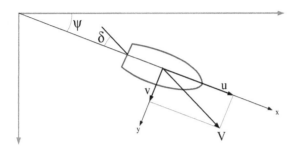

Figure 4. Coordinates and notation used to described the equations

Let 'V' be the total velocity, 'u' and 'v' the x and y components of the velocity, and 'r' the angular velocity of the ship. In normal steering the ship makes small deviations from straight-line course. The natural state variables are the sway velocity 'v', the turning rate 'r', and the heading 'ψ'. The equations 2 are obtained, where 'u' is the constant forward velocity, 'l' the length of the ship and 'a' and 'b' are parameters of the model ship.

$$
\begin{aligned}
\frac{dv}{dt} &= \frac{u}{l}a_{11} + ua_{12}r + \frac{u^2}{l}b_1\delta \\
\frac{dr}{dt} &= \frac{u}{l^2}a_{21}v + \frac{u}{l}a_{22}r + \frac{u^2}{l^2}b_2\delta \\
\frac{d\psi}{dt} &= r
\end{aligned}
\tag{2}
$$

From equation 2 is determinated the transfer function from rudder angle to heading in the equation 3.

$$
G(s) = \frac{K(1+sT_3)}{s(1+sT_1)(1+sT_2)}
\tag{3}
$$

where,

$$
\begin{aligned}
K &= K_0 u/l \\
T_i &= T_{i0}l/u \qquad i = 1, 2, 3
\end{aligned}
\tag{4}
$$

The parameters K_0 and T_{i0} are parameters of ship model. In many cases the model can be simplified to equation 5.

$$
G(s) = \frac{b}{s(s+a)}
\tag{5}
$$

where,

$$b = b_0 \left(\tfrac{u}{l}\right)^2 = b_2 \left(\tfrac{u}{l}\right)^2$$
$$a = a_0 \left(\tfrac{u}{l}\right)$$

(6)

This model is called the first order Nomoto model of a ship. Its gain 'b' can be expressed approximately as expression 7.

$$b = c \left(\frac{u}{l}\right)^2 \left(\frac{Al}{D}\right)$$

(7)

where $'D'$ is the displacement (in m^3), $'A'$ is the rudder area (in m^2) and $'c'$ is a parameter whose empirical value is approximately 0.5. The parameter $'a'$ will depend on trim, speed and loading and its sing may change with the operating conditions.

6. System used to verify the proposed method

To illustrate the method proposed in this document for the automatic steering of ships, it is going to be applied to a freighter with 161 meters of length, which displacement will range from 8,000 m^3 in the vacuum until the 20,000 m^3 full load. The velocity at which will be able to navigate will be more than 2 meters per second, for which is perfectly valid the model used, to 8 meters per second maximum velocity. It is necessary to specify that the Servo-rudder operates at a speed of 4 meters per second limited to ± 30 degrees, according to this description the model shown in Figure 5.

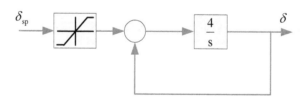

Figure 5. Servo-rudder blocks diagram

The transfer function of the freighter remains as it is indicated in the expression 8, in which was only given the value to $'c'$ that is 0.5.

$$G(s) = \frac{0.5 \left(\frac{u}{l}\right)^2 \frac{Al}{D}}{s \left(s + a_0 \frac{u}{l}\right)}$$

(8)

It will be replaced the fixed values commented, where it should be pointed out for the case of the freighter, that the parameter a_0 has a value of 0.19 and the variable will be replaced in each case as necessary.

To carry out the simulations and to achieve the desired data for implementing the proposed model, it is used Matlab/Simulink. For which it is edited in the first place the following control scheme based on descriptions made previously, shown in Figure 6.

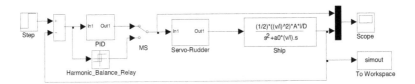

Figure 6. Control scheme in Simulink format

The PID block houses, the scheme shown in Figure 7, whose structure is as the one explained at the beginning of the document, in which it should be noted for a better approximation to the real system, the congestion at the output of the block adder.

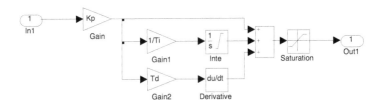

Figure 7. PID block in Simulink format

The servo-rudder block has inside, the diagram of figure 8, which is neither more nor less than the diagram of the Servo of Figure 5 in Simulink format.

6.1. System operation conditions

The operation conditions of the system are infinite therefore certain values have to be chosen. It makes no sense to obtain parameters for multiple cases so it is necessary to make a coherent estimate to achieve good results. One approach is to choose a reasonable amount of equidistant values and observe the changes of the parameters for each case. If there are substantial changes from one value to another then an opportunity of taking new intermediate values between them will be provided.

There are own terms of the ship that will not vary as can be the area of the rudder. In this case, the only term that will define the operation conditions or gains adjustment rules of the

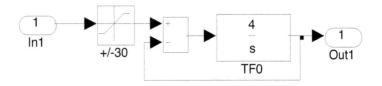

Figure 8. Servo-rudder in Simulink format

controller, are the displacement and the velocity of the ship. It is necessary to highlight that the displacement is not going to be obviously a property that is constantly changing as can do the velocity, even if it is in a slow way. Taking into account the above and the ranges of values that can take each of the two parameters from which the model depends on, it is established the Table 1 of possible conditions.

D	u
	2
8000	4
	6
	8
	2
12000	4
	6
	8
	2
16000	4
	6
	8
	2
20000	4
	6
	8

Table 1. Working points selected

6.2. Obtaining the controller parameters for each operation status

In the stage of obtaining the parameters of different working points, in the control implementation by the scheduling of the virtual controller instead of PID controller in parallel could be selected a hysteresis block. This is an attempt to obtain the controller parameters using the Relay Feedback method and it will be discussed in a summarized form.

6.2.1. Relay-feedback method

This is an alternative way to the chain closed method of Ziegler-Nichols [24–28], for the empirical location of the critical gain (K_c) and the period of sustained oscillation (T_c) of the system. It uses of the method of relay (Relay Feedback) developed by Aström and Hägglud [13, 29], which consists in leading the system to the oscillation state by the addition of a relay as it is shown in Figure 9.

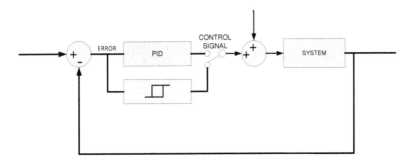

Figure 9. Application scheme of Relay-Feedback

This oscillation taken from the system has a period with approximately the same value as the period of sustained oscillation Tc (critical period). In the experiment it is recommended to use a relay with hysteresis which characteristics like the one shown in Figure 10 with an amplitude d and a width of the hysteresis window h.

After the assembly is done, it will proceed as follows to get the parameters mentioned:

1. Leading the process to put the process in steady state, with the system regulated by the PID controller, with any parameters that let us achieve that status. It will be taken note of the control signal values and the output of the process in those conditions.

2. Then the control is finished with the relay, instead of the PID controller. As a set point it is given the value read in the output of the process in the previous step. It is introduced in the input shown in Figure 7 as the Offset, the control signal value taken in the previous paragraph which is necessary to put the process in steady state.

3. The process is situated into operation with the indications made in the previous paragraph, and it is expected to become regular in the output (in practice it can be considered to have reached this state when the maximum value of the output repeats the same value for at least two consecutive periods).

4. It will be noted down the two parameters shown in Figure 11, where T_c is the sustained oscillation period.

5. The critical gain of the process is determined by expression 9.

$$K_c = \frac{4d}{\pi\sqrt{a^2 - h^2}} \tag{9}$$

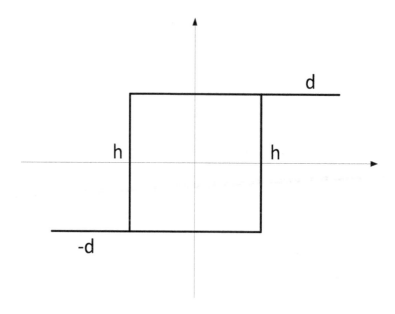

Figure 10. Hysteresis for Relay-Feedback

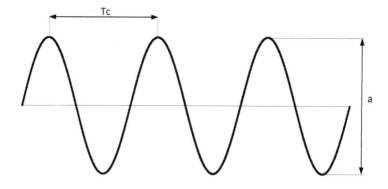

Figure 11. Parameters to read in the sustained oscillation (critic period)

The Relay Feedback has the advantage that adjustment can be made on the set point and it can be carried out at any time. However, a problem is that to tune, the process must exceed the set point on several occasions and there could be cases in which this is inadvisable because of the damage they can cause during the process.

6.2.2. Obtaining the parameters T_c and K_c

In the particular case shown in this paper, there is no need to implement the hysteresis mentioned in the explanation of Relay Feedback with a window, because works with a relatively slow system. Instead, a simple comparator as the one shown in Figure 12 will be sufficient.

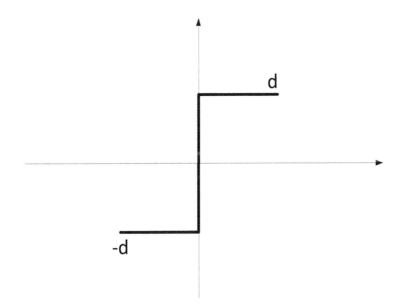

Figure 12. Hysteresis for the plant tested

Logically the value of h is zero and the value of d is 0.5. It is established as set point a value of 0.5, and the offset for this case is not necessary because it would be zero. Under these conditions the system becomes operational, and the result obtained is shown in figure 13.

It is necessary to pay attention to the final zone, where the oscillation is now stabilized and periodic, and with the expressions commented above for the Relay Feedback method, the extracted parameters are T_c and K_c.

6.2.3. Obtaining the PID controller initial parameters

With the parameters that have been obtained in the previous paragraph, it is possible to get the controller parameters applying direct formulas, achieving the three terms of the regulator.

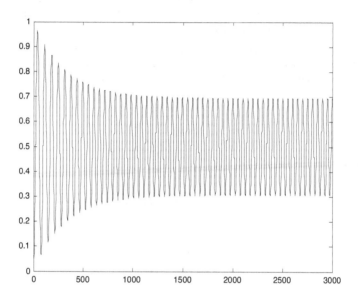

Figure 13. Result of application of Relay-Feedback in a working point

K	$=0.6 \times K_c$
T_i	$=0.5 \times T_c$
T_d	$=0.125 \times k_c$

Table 2. Ziegler-Nichols formulas for closed chain

In this case it will be necessary to obtain them for a criterion of changes in the load (for load disturbances rejection).

Taking this into account the expressions to tune controllers in closed chain of Ziegler-Nichols will be applied. They are the pioneer formulas for obtaining controller parameters, and they are good at changes in the load. The specification that it is trying to obtain is a list of overshoot of a quarter decay ratio, which means that in the face of the input of a disturbance, the successive overtopping of reference, are four times lower than the previous one (damping factor of 1/4). Such expressions are shown in Table 2.

6.2.4. Fine tuning of the controller

The parameters obtained in the preceding paragraph would be necessary subject to a fine-tuning, because the results reached are not suitable.

In this most delicate task of adjustment it is necessary to indicate that it should not saturate the output controller at any time. It is necessary to reach a compromise, since an excess of proportional gain causes a fast response in the output and with little overshoot, which apparently is ideal, but under these conditions the servo is constantly fluctuating, which will cause it to deteriorate in a short period of time. As conclusion, it is going to search gradual outputs, without saturation or sudden changes such as the case shown in Figure 14.

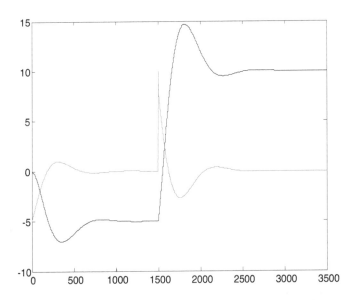

Figure 14. Example of steering and control signal to the rudder

It is necessary to indicate that for this model the method of adjusting parameters Ziegler-Nichols is not the ideal, since the initial values of the parameters do not give good results, differing greatly from those achieved after fine-tuning.

6.2.5. Parameters obtained for each case

Taking all the comments above into account all the comments above we obtain the controller parameters fine-tuned for each of the cases discussed above, which seeks a criterion of minimum overshoot and maximum speed for the restrictions presented in the preceding paragraphs. In this way we reach the parameters of the table 3.

D	u	K	T_i	T_d
	2	3	350	90
8000	4	1	300	80
	6	1	350	80
	8	1	350	80
	2	4	350	80
12000	4	3	300	90
	6	1	350	95
	8	1	350	85
	2	6	400	90
16000	4	3	300	90
	6	1	350	90
	8	1	300	85
	2	6	450	90
20000	4	4	350	90
	6	1	350	90
	8	1	300	90

Table 3. Controller parameters obtained for each rule

6.3. Implementation of the neural network

It has been designed a neural network type MLP (Multi Layer Perceptron) for scheduling of each one of the controller constants K, T_i and T_d. They all have two inputs, which are the displacement and the velocity of the ship, and one output that is the corresponding constant. The neuronal network has an intermediate layer with 5 neurons for K and 6 for T_i and T_d. This structure has been adopted after many tests with different numbers of neurons in the middle layer (tests were made from 4 to 9 neurons in the middle layer) for each of the neural networks. The activation functions of neurons in the middle layer are a kind of hyperbolic tangent, except in the output layer, where one neuron is with a linear function.

Once this configuration is selected it is shown the different characteristics of the training carried out with backpropagation learning. It has been made the training of K, T_i and T_d at 531, 705 and 686 respectively epochs, with an average error at the end of the training less than 1%. The artificial neural networks have been trained off-line, although the checking of its proper operation has been performed on line.

7. System assembly and verification of results

It is implemented in Simulink (Figure 15) the system diagram. If it is compared with the model used for obtaining the controller parameters in the different representative points of work, it is possible to observe that the relay block , has been eliminated, and the PID controller has been replaced for a block, which name is Neuro-PID.

The internal diagram blocks of the Neuro-PID controller, is shown in Figure 16 in the format that has been implemented in Simulink.

As it is shown in Figure 16 a PID controller is implemented where the earnings are the outputs of the artificial neural networks and which inputs are the displacement and the

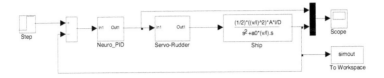

Figure 15. System on Simulink format

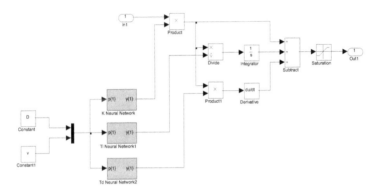

Figure 16. Neuro-PID Block

velocity of the ship. The pins In1 and Out1 are the Neuro_PID block pins of the figure 15, which control signal is joined directly to the servo-rudder.

By this way, the implemented controller will choose the most appropriate parameters for the area in which it is working. It should be noted that more points could be obtained to train neural networks, but it would be more expensive. Furthermore, the neural network itself follows the tendency of data, already interpolates properly between them, showing one of the advantages of its use.

To validate the model created it is resorted to its simulation with different values of the parameters, on which depend the velocity model and displacement.

It has been made different tests at multiple points of work, and in Figure 17 is shown four representative examples in which in all cases is made a steering to -5ž and once stabilized to +10ž.The answer is satisfactory and similar in all cases with the only difference in velocity due to the different velocities. It should be noted that at small velocities to get an adequate response of the steering, similar to the one of the entire range, it is necessary to saturate the

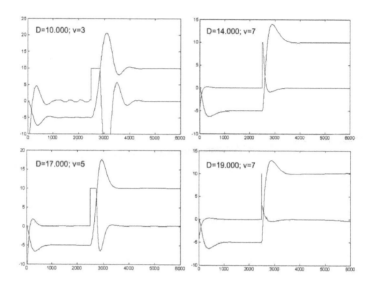

Figure 17. Model response to different operating conditions

output of the controller, fact unwanted at the time of fine-tuning, but necessary to maintain the specifications within a range of values.

After the results achieved, it is pointed a satisfactory behaviour of the implemented system, in which the desired results of uniformity are achieved in the operation, regardless of conditions, from which depends the model of the ship.

8. Conclusion

Obviously, in non-linear systems, such as the case of the steering a ship studied in this document, and also working across a wide range of operation, and that could be divided in zones with a linear behaviour, in which the control is also feasible using a type PID controller, the option of its use with the method proposed in this paper is an option to take into account.

As an alternative to the different types of autotuning PID's one of the easiest solutions is the one developed in this article. It is necessary to accentuate that it is not an easy solution to adopt, especially with continuous controllers, but with the addition of programmable control devices this labour becomes comparatively simple.

Emphasize that difficulties in the use of PID controllers working with Gain Scheduling, have the problem of taking those points which are significant, interpolation between them and also could happen that the system is stable at selected points but not between them. With the use of artificial neural networks all these drawbacks are softened to a large extent, since all of them are solved with the use of this aspect of artificial intelligence.

This methodology provides a uniform response of the system throughout the whole operating range of the ship, regardless of displacement, or the velocity, parameters of which the model depends on. If this depends on other factors, the methodology could be applied equally to them.

Author details

José Luis Calvo Rolle[1] and Héctor Quintián Pardo[2]

[1]University of Coruña, Spain
[2]University of Salamanca, Spain

References

[1] G. C. Nunes, A. A. Rodrigues Coelho, R. Rodrigues Sumar, and R. I. Goytia Mejía. A practical strategy for controlling flow oscillations in surge tanks. *Latin American applied research*, 37:195–200, 07 2007.

[2] O. Begovich, V. M. Ruiz, G. Besancon, C. I. Aldana, and D. Georges. Predictive control with constraints of a multi-pool irrigation canal prototype. *Latin American applied research*, 37:177–185, 07 2007.

[3] David A. Mindell. *Between Human and Machine: Feedback, Control, and Computing Before Cybernetics*. Number xiv, 439 p in Johns Hopkins studies in the history of technology. The Johns Hopkins University Press, Baltimore, 2002.

[4] S. Bennett. Nicholas minorsky and the automatic steering of ships. *Control Systems Magazine, IEEE*, 4(4):10 –15, november 1984.

[5] M.H. Moradi. New techniques for pid controller design. In *Control Applications, 2003. CCA 2003. Proceedings of 2003 IEEE Conference on*, volume 2, pages 903 – 908 vol.2, june 2003.

[6] Karl Johan Åström and Tore Hägglund. *PID Controllers: Theory, Design, and Tuning, 2nd Edition*. ISA, 1995.

[7] Y. Li, W. Feng, K.C. Tan, X.K. Zhu, X. Guan, and K.H. Ang. Pideasy(tm) and automated generation of optimal pid controllers. In *Proc. Third Asia-Pacific Conference on Measurement and Control*, pages 29–33, Dunhuang, China, Sept 1998.

[8] K. Nomoto and K. Taguchi. On the steering qualities of ships (2). *Journal of the Society of Naval Architects of Japan*, 101:n.p., 1957.

[9] N.H. Norrbin. On the design and analyses of the zig-zag test on bases of quasi to line frequency response. Technical Report 104-3, The Swendish State Experimental Shipbuilding Tank (SSPA), Gothenburg, Sweden, 1963.

[10] Bech.M.I. and L. W. Smith. Analogue simulation of ship manoeuvers. Technical Report Hy-14, Hydro and Aerodynamics Laboratory, Lyngby, Denmark.

[11] S.K. Bhattacharyya and M.R. Haddara. Parametric identification for nonlinear ship maneuvering. *Journal of Ship Research*, 50(3):197–207, 2006.

[12] M. H. Casado, R. Ferreiro, and F. J. Velasco. Identification of nonlinear ship model parameters based on the turning circle test. *Journal of Ship Research*, 51(2):174–181, 2007.

[13] Karl Johan Åström and Bjorn Wittenmark. *Adaptive Control*. Addison-Wesley Longman Publishing Co., Inc., Boston, MA, USA, 2nd edition, 1994.

[14] Karl Johan Åström and Hägglund Tore. The future of pid control. *Control Engineering Practice*, 9(11):1163 – 1175, 2001. <ce:title>PID Control</ce:title>.

[15] Eduardo F. Camacho and Carlos A. Bordons. *Model Predictive Control in the Process Industry*. Springer-Verlag New York, Inc., Secaucus, NJ, USA, 1997.

[16] W.J. Rugh. Analytical framework for gain scheduling. *Control Systems, IEEE*, 11(1):79 –84, jan. 1991.

[17] B. Clement and G. Duc. An interpolation method for gain-scheduling. In *Decision and Control, 2001. Proceedings of the 40th IEEE Conference on*, volume 2, pages 1310 –1315 vol.2, 2001.

[18] W.M. Lu, K. Zhou, and J.C. Doyle. Stabilization of lft systems. In *Decision and Control, 1991., Proceedings of the 30th IEEE Conference on*, pages 1239 –1244 vol.2, dec 1991.

[19] K. Hiramoto. Active gain scheduling: A collaborative control strategy between lpv plants and gain scheduling controllers. In *Control Applications, 2007. CCA 2007. IEEE International Conference on*, pages 385 –390, oct. 2007.

[20] Joo-Siong Chai, Shaohua Tan, and Chang-Chieh Hang. Gain scheduling control of nonlinear plant using rbf neural network. In *Intelligent Control, 1996., Proceedings of the 1996 IEEE International Symposium on*, pages 502 –507, sep 1996.

[21] Jin-Tsong Jeng and Tsu-Tian Lee. A neural gain scheduling network controller for nonholonomic systems. *Systems, Man and Cybernetics, Part A: Systems and Humans, IEEE Transactions on*, 29(6):654 –661, nov 1999.

[22] Chian-Song Chiu, Kuang-Yow Lian, and P. Liu. Fuzzy gain scheduling for parallel parking a car-like robot. *Control Systems Technology, IEEE Transactions on*, 13(6):1084 – 1092, nov. 2005.

[23] E. Applebaum. Fuzzy gain scheduling for flutter suppression in unmanned aerial vehicles. In *Fuzzy Information Processing Society, 2003. NAFIPS 2003. 22nd International Conference of the North American*, pages 323 – 328, july 2003.

[24] M. Zhuang and D.P. Atherton. Tuning pid controllers with integral performance criteria. In *Control 1991. Control '91., International Conference on*, pages 481 –486 vol.1, mar 1991.

[25] P. Cominos and N. Munro. Pid controllers: recent tuning methods and design to specification. *Control Theory and Applications, IEE Proceedings -*, 149(1):46 –53, jan 2002.

[26] Karl Johan Åström, H. Panagopoulos, and Tore Hägglund. Design of pid controllers based on non-convex optimization. *Automatica*, 34(5):585 – 601, 1998.

[27] Tore Hägglund and Karl Johan Åström. Revisiting the ziegler-nichols tuning rules for pi control. *Asian Journal of Control*, 4(4):364–380, 2002.

[28] J. G. Ziegler and N. B. Nichols. Optimum settings for automatic controllers. *Journal of Dynamic Systems, Measurement, and Control*, 115(2B):220–222, 1993.

[29] Karl Johan Åström and Tore Hägglund. Automatic tuning of simple regulators with specifications on phase and amplitude margins. *Automatica*, 20(5):645 – 651, 1984.

Heuristics for User Interface Design in the Context of Cognitive Styles of Learning and Attention Deficit Disorder

Sandra Rodrigues Sarro Boarati,
Cecilia Sosa Arias Peixoto and
Cleberson Eugenio Forte

Additional information is available at the end of the chapter

1. Introduction

Several companies and institutions now realize knowledge as an active relevant for the market organization differentiation. This scenario explains the need for systems that assist the user in the acquisition process and knowledge management. Intelligent systems, known as expert systems (ES) [19] serve to this purpose in the extent that they have signed as facilitators in this process. These are systems that are based on expert knowledge, on any subject, in order to emulate human expertise in the specific field. To obtain this knowledge, the knowledge engineers, also called software engineers, need to develop methodologies for intelligent systems. In this area there is still no unified methodology that provides effective methods, notations and tools to aid in development. Among the most used technologies we can mention: KADS [10], MIKE [2] and Protégé [8]. KADS is a structured way of developing these systems that provides a special focus on the characteristics and problems of development of the SE. KADS uses the waterfall model as a basis to guide the development and adds refinement stages of use and knowledge [10]. Moreover, the MIKE methodology (Model-Based and Incremental Knowledge Engineering) makes use of formal specification and semi-formal techniques during the incremental development of the system. The phases of this model are four (Figure 1), being the first, knowledge acquisition, made in a cyclic manner between the subphases of task analysis, model construction and evolution. After the acquisition phase, the design, implementation and evolution are cyclical until the system is built.

It is an approach to build domain ontologies and includes the notion of a library of reusable problem-solving methods (PSMs) that perform tasks. In PROTÉGÉ-II, PSM are decomposable into subtasks. Other methods, sometimes called sub methods, can perform these subtasks. Primitive methods that cannot be decomposed further are called mechanisms. This decomposition is made to allow the reuse of knowledge, an essential part of the methodology. The choice of which method is suitable for the development of an intelligent system is essential, given the complexity of systems in the field of artificial intelligence (AI). It is important to define a process that systematizes the life cycle, allowing a greater skill in eliciting and models description. In this work, it was elaborated a guide for knowledge acquisition based on ontologies [9] and applied to the extension of an expert system of recommendations (guidelines) for designing human-computer interfaces.

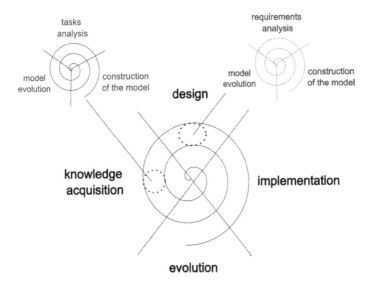

Figure 1. Stages of knowledge acquisition, design, implementation and evolution from MIKE.

This expert system called GuideExpert was expanded to include recommendations about profiles of users with learning disorder (TA), attention deficit disorder / hyperactivity (ADD / H) and advices on cognitive learning styles (ECAS). The learning disorder is defined where individuals can not develop as expected in appropriate age scholl [22], on the other hand the deficit of attention disorder/ hyperactivity and impulsivity [1]. In turn, the cognitive learning styles represent a categorization of the cognition particularities with their respective skills [21]. There are several recommendations on how computer interfaces should be designed in order to attend, in a satisfactory way, users with learning disorders and different cognitive styles, among other features. Thus, the aim of incorporating this knowledge to the GuideExpert base needed the establishment of a process for knowledge acquisition which will be presented in the next section.

2. Knowledge acquisition for the new GuideExpert acquaintance

The GuideExpert system is an expert system developed to assist the designer of human-computer interfaces during the phases of design and evaluation of interfaces. The system consists of five elements: user interface, inference engine, working memory, knowledge base and database. Figure 2 shows the architecture of the expert system. Through a series of questions and screens, the system selects a series of recommendations (called guidelines) of experts in the field of interfaces. The GuideExpert in its first version consisted of three hundred and twenty six guidelines of elicited interfaces projects from various authors and experts in the field of interfaces. To search the knowledge base of the GuideExpert it was defined the meta-guidelines. They are concepts which embrace the guidelines according to the common characteristics and goals [6].

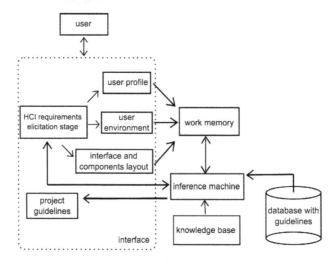

Figure 2. GuideExpert expert system

In the second phase of the project, it was seen the need to incorporate in the system, the guidelines relative to the diversity of user profiles. We identified several recommendations, heuristic and knowledge about adults, children, handicapped users, users with deficits of attention and etc. In order to make the knowledge acquisition in this domain it was elaborated a guide based on Ontologies. Ontology is a formal and explicit specification of a shared conceptualization [9]. The ontologies are used to structure and share the knowledge. They can be seen as the highest level in a hierarchy of knowledge composed of vocabularies, thesauri, taxonomies, ontologies and frames. A taxonomy is to classify information in a hierarchy (tree) with the generalization relationship "kind-of" (parent-child) [4].

The existence of a taxonomy in GuideExpert system, formed by the meta-guidelines, motivated the ontology conceptualization for projects in human-computer interfaces. For the creation of ontologies there are different methods belonging to the Ontological Engineering area. The goal of these methods is to provide tasks or steps to be followed in creating the ontology. Among the best known in litaratura we can mention: the method of Uschold [23], Horrocks [11], and Noy and McGuiness [14]. Uschold proposal called "Skeletal Methodology" uses scenarios to describe knowledge. Questions or types of questions are made primarily in order to specify the knowledge that is not being adequately addressed by the ontologies and that will be conceptualized through questions.

The Ian Horrocks method, is composed of the following phases:

• Determine how the world (domain) must function

• Determine domain classes and properties

• Determine domains and scopes (range) for this domain property

• Determine classes characteristics

• Add individual and relationships

• Iterate the steps until end the ontology conceptualization

• Specify an ontology

• Care if the ontology already exists

• Verify the consistency using a rationalization tool or inference motor

• Verify if classes are coherent

The Ian Horrocks method goes through the definition domain, the definition of classes, properties, etc. Another methods also start the creation of ontologies focusing on domains, classes and hierarchies, but though questionings. This is the case of the 101 method proposed by Natalya Noy and Mac Deborah McGuiness [14]. This method brings together recommendations and experiences using the tool for editing ontologies, Protégé 2000 [17], the Ontolingua language [16] and the tool Chimaera [5]. This methodology focuses on the Ontology conceptualization phase, divided into seven steps, which involves: definition of the ontology classes, storage of the classes in the hierarchy, defining properties and definition of the instances.

The steps that comprise the methodology proposed by Noy and McGuinness are [14]:

1. Determine the scope of the ontology: to determine the domain and scope it is suggested the following questions:

1.1 Which domain is desired to cover with the Ontology?

1.2 With which purpose the Ontology will be used?

1.3 For wich questions the Ontology mus provide answers?

1.4 Who is going to use and maintain the ontology?

2. Consider the reusing of another ontologies: is whether ontology exists and refine or extend the model to new domain or task.

3. Enumerate important terms of the ontology: is to define a list of the most common terms in the domain and the properties they possess:

3.1 Which are the terms that are desired to be included?

3.2 Which are the properties of these terms?

4. Define the classes and class hierarchy: is to determine the consistency of the class-subclass hierarchy, ie it should be noted that a class has more or less subclasses There are several strategies for defining the hierarchy: top-down, bottom-up or mixed.

5. Set the properties of the classes: it is to create some concepts in the hierarchy, and then their properties.

6. Set the values of properties (also called facets) is to describe the types of data (values), allowed values, domain, scope, minimum and maximum number (cardinality) for property values, and others.

7. Create instances: it consists in choosing the class for which you want to create instances filling the property values for each instance.

The method of Noy and McGuiness was adopted here in order to define concepts, properties and relationships for the domain knowledge of Human Computer Interface. Thus, we defined the classes and relationships of the knowledge of the system GuideExpert.

The method of Noy and McGuiness was adopted here in order to define concepts, properties and relationships for the domain knowledge of Human Computer Interface. Thus, we defined the classes and relationships of the knowledge of the system GuideExpert.

2.1. GuideExpert Ontology

The Guidelines of the project of the GuideExpert were classified in a taxonomy consisting of 10 meta-guidelines, including: Feedback System, Data Protection, Documentation, interaction, presentation of data, Internationalization of interfaces, Colors, Terminology Interface, Design, and Assisting people with disabilities.

From this taxonomy, the ontology was defined, following the steps of the methodology proposed by Noy and McGuinness [14]. At the stage of conceptualization of theontology for the system GuideExpert it was elicited domain, objective information, users, tasks and resources, namely:

• Domain definition: concepts of human-computer interfaces;

• Definition of the main goal or purpose: create an ontology to guide the search for guidelines (recommendations) of project interface;

• Definition of informations about the ones that should provide answers: related to the concepts established on the metaguidelines.

- Definitions of the users: the potencial users of the ontologies are the engineers of the interfaces Human-Computer and learners of the development of interfaces projects.

- Definition of tasks: the main tasks held for the development of the ontologies consist in following the steps of the Noy and McGuiness methodology in this first step of the research.

- Definition of the resources: the resources needed for the development of the ontology are the computational tool to model the ontology and the language to formalize it. These steps will be developed in the next steps of the project.

Subsequently, based on the taxonomy, it was defined the classes and relationships. The main classes were expanded in most relationships and subclasses according to the recommendations of guidelines that exist in the system GuideExpert. Figure 3 shows the concept of class interaction with the relationships used (Table 1).

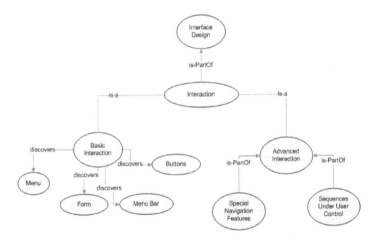

Figure 3. Diagram of the subclass interaction ontology.

Relationship	Meaning
Is-a	Indicates that one class is subclass of another.
s_PartOf	Indicates that one classe is part of another
hasValue	Indicates that one class obtain values from another class.
defines	Indicates that one class defines a determinate concept from another class
discovers	Indicates that a class discovers a determinate pattern by another class

Table 1. relationships dictionary

Once elicited the ontology, it was used for the acquisition of knowledge about the guidelines for user profiles. For each of the ontology classes it were surveyed authors and experts in the field of interfaces. Thus, it was possible to expand the taxonomy with more concepts and relationships. The method of eliciting followed, based on the ontology, was the following (Figure 4):

1.   Selecting of a class of the ontology of the GuideExpert.

2.   Elicit questions that can be answered by specialists.

3.   Make the acquisition of this knoweledge.

4.   If a new knoweledge if found, describe the class and add it to the ontology.

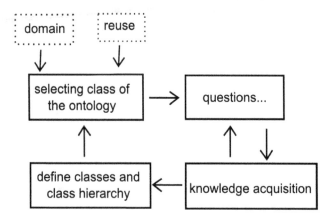

Figure 4. Knowledge acquisition process based on ontologies used in GuideExpert.

It was incorporated into the knowledge base of the GuideExpert, one hundred thirty six specific guidelines of the profile of users and special features such as: ADHD users, users with Analog Analytical cognitive style, etc.

3. Heuristics for user interface design in the context of cognitive styles of learning and attention deficit disorder

The user population is not a system composed of only one type of user. In general, there is a mixture of multiple profiles of users who need to somehow get their needs met [15].

Speaking of users interacting with computers, we refer to the user's knowledge that should be taken into account in the design of an IHC. Below are some features that must be observed during the interface design [12],[13],[20]:

• The presence of an internationalized system or used in more than one country or region. Each country or region has its own peculiarities. Dialects, cultures, ethnicities, races, etc. All these elements end up generating needs that to be satisfied;

• These characteristics are considered common in a traditional interface, may not correspond to those made for children. They have unique needs for their age. Beyond these specific needs for the children users the designers need to deal with the dangers that are usually present in a web environment, such as pornography and violent or racist content;

• The existence of elderly among the users should be checked for these and needs met;

People with special needs are another installment of the user community of the system and need adjustments in the system to operate in the environment without difficulty.

For the ECAs we used the basis of the research article: Project Tapejara from Souto [21]. The ECAs refer the subject's characteristic way of learning new concepts or even to generate elaborations of prior knowledge. According to Madeira, et al. (Bica et al., 2001), the ECAs are: Analogue-Analytical (AA), Concrete-Generic (CG), Deductive-Evaluative (DA), Relational-Synthetic (RS) and Synthetic-Evaluative (SA).

• Analogue-Analytical Style: using prior knowledge, seek information using standards of comparison. The information is analyzed in blocks. Performs elaborations relating the previously acquired knowledge and the new.

• Concrete-Generic Style: trying to understand the contents in a linear and sequential way; works with memorization through systematic exemplification. The individual is pragmatic and careful.

• Deductive-Evaluative Style: is systematic and critical, making analysis of the information. Does not consider concrtet examples. His work and attention are high.

• Relational-Synthetic Style: the individual better understands the information through pictures, colors, diagrams, etc. Has the ability to abstract hypotheses.

• Synthetic-Evaluative Style: by intercalation between the global view of data and its evaluation, seek to learn new information, analyzing them as a whole. They organize the study material, preferring theoretical material; they are systematic.

Souto [21] states that users having Analytic-Analog style "may require more time for learning, because when confronted with new information, tend to get a considerable depth on the subject, through intense".

Users having Concrete-Generic style, Souto [21] says that they "tend to be pragmatic and careful in their learning situation. The learning objectives, evaluation criteria and feedback must be clear to this style, because then he can work towards the goals".

Users having Deductive-Evaluative style "may come to disregard large number of concrete examples, when they believe they have already understood the logical pattern underlying the new information" [21].

Relational-Synthetic style users "tend to have ease of mind to work with images and appreciate the use of charts, diagrams and demonstrations. They are profecient in working with charts and mind maps" [21].

Synthetic-Evaluative style was not evaluated because "the subject of this class carryies features from the analytics, synthetics, and evaluative" [21].

In analysis of the cognitive styles of learning, this study concluded that:

* Analogue-Analytical style (AA): the user cares about the information in blocks, prefers everything organized. Regarding to the interfaces, it is more preferred that objective, organized, with infomation presented clearly, without many shades. These users prefer the basic colors, or colors worked according to the information and its importance.

* Concrete-Generic style (CG): colors should be worked upon the variations of the same color, ie, linearly, less elaborate, meeting the expectations of those who receive information and belongs to that kind of style.

* Deductive-Evaluative style (DE): the information may be indirect, can cause the user having to think to acquire information.

* Relational-Synthetic style (RS): users better understand the information presented through images, different colors, diagrams, etc. The own style is defined with the use of characters, colors and pictoric examples.

* Synthetic-Evaluative style (SE): the presentation is not what counts most, but the content. A presentation with plenty of written information is of utmost importance. Coherence and justification of the data determine how the user will engage with the text.

From the CLSs (Cognitive Learning Styles) studied, we found that it is extremely important to take into account the characteristics of each user in the construction of new interfaces for the computer. We must take into consideration the colors used in the construction of the interface, the way that the information must be displayed, directly or indirectly, if the interface must be objective or not, using figures. It was necessary to relate the CLSs with users suffering from Attention Deficit Disorder (ADD) or Attention Deficit Disorder with Hyperactivity (ADHD) because they are related to learning style and how each acquires knowledge.

The people with ADD and ADHD can not develop the scholar knowledge as expected for their ages. The diagnostic in the scholar age is common because in this period can be found the difficulties of attention and remain silent as the studies of Siqueira and Gusgel Giannetti [22], Rosa Neto and Poeta [18].

This research contributed to the GuideExpert tool incorporating new knowledge items to it, enabling the groups of users with different learning styles (CLSs) and users suffering from attention deficit disorder (ADD and ADHD) to obtain special guidelines.

Meta*GUIDELINES*	
9.13	Volume/sound design
9.14	Mouse design
9.15	Keyboard design
9.16	Source design
9.17	Help design
9.18	Link design
9.19	Figure/image design
10	Assistance to people with disabilities
10.3	Asssitance to people having ADHD
10.4	Assistance to people having visual disabilities
11	User Types
11.1	Elderly Users
11.2	Children Users
12	CLS – Cognitive Learning Style
12.1	AA – (Analogue-Analytical)
12.2	DE – (Deductive-Evaluative)
12.3	RS – (Relational-Synthetic)
12.4	CG – (Concrete-Generic)

Table 2. Augmented expert system taxonomy

```
R1: When carriers_ADD == child
    Then meta-guideline = help_add; user_child
R2: When carriers_ADD == elderly
    Then meta-guideline = help_add; user_elderly
```

Figure 5. Rule Knowledge Base - People with ADD - Children – Elderly

Eighteen new classes were created and added to the taxonomy, according to the surveys [3], [22] and [18]. The newly added metaguidelines are listed at Table 2

R1: When carriers _ADD == child
Then meta-guideline = help_add; user_child
R2: When carriers _ADD == elderly
Then meta-guideline = help_add; user_elderly
R3: When carriers _ADHD == child

R1: When carriers _ADD == child
Then meta-guideline = help_adhd; user_child
R4: When carriers _ADHD == elderly
Then meta-guideline = help_adhd; user_elderly
R5: When carriers _ colorblindness== child
Then meta-guideline = help_ colorblindness; user_child
R6: When carriers _ colorblindness == elderly
Then meta-guideline = help_ colorblindness; user_elderly
R7: When carriers_visual_impairment == child
Then meta-guideline = help_ visual_impairment; user_child
R8: When carriers _ visual_impairment == elderly
Then meta-guideline = help_ visual_impairment; user_elderly
R9: When carriers_special_need == child
Then meta-guideline = help_ special_need; user_child
R10: When carriers_ special_need == elderly
Then meta-guideline = help_special_need; user_elderly
R11: When eca_aa == child
Then meta-guideline = eca_aa; user_child
R12: When eca_aa == elderly
Then meta-guideline = eca_aa; user_elderly
R13: When eca_cg == child
Then meta-guideline = eca_cg; user_child
R14: When eca_cg == elderly
Then meta-guideline = eca_cg; user_elderly
R15: When eca_da == child
Then meta-guideline = eca_da; user_child
R16: When eca_da == elderly
Then meta-guideline = eca_da; user_elderly
R17: When eca_rs == child
Then meta-guideline = eca_rs; user_child
R18: When eca_rs == elderly
Then meta-guideline = eca_rs; user_elderly

Table 3. Rule Selection

The knowledge basis of GuideExpert consist in the "WHEN-THEN" rules. This study adds

the 18 selected rules that were created according to the preview research to the base already

built, and follow the same syntax as shown in Table 3.

GUIDELINES
1. Guideline: Use blinking displays 2-4 Hz with great care and in limited areas [1].
2. Guideline: Use up to three sources to draw attention [1].
3. Guideline: Use the inverse staining [1].
4. Guideline: Use up to four color standards [1].
5. Guideline: Use only two levels of intensity [1].
6. Guideline: Children approve the use of animations and sound [2].
7. Guideline: Avoid using scrolling for children [2]

REFERENCES
[1] SHNEIDERMAN, B. **Designing the User Interface:** Strategies for Effective
Human-computer Interaction. 3. ed. Boston: Addison Wesley Longman, Inc., 1998.

[2] NIELSEN, J. **Children's Websites:** Usability Issues in Designing for Kids.
Alertbox: september 13, 2010. Disponível em: <http://www.useit.com/alertbox/
children.html>. Acesso em: 24 nov. 2011.

Figure 6. Guidelines - People with ADD - Children

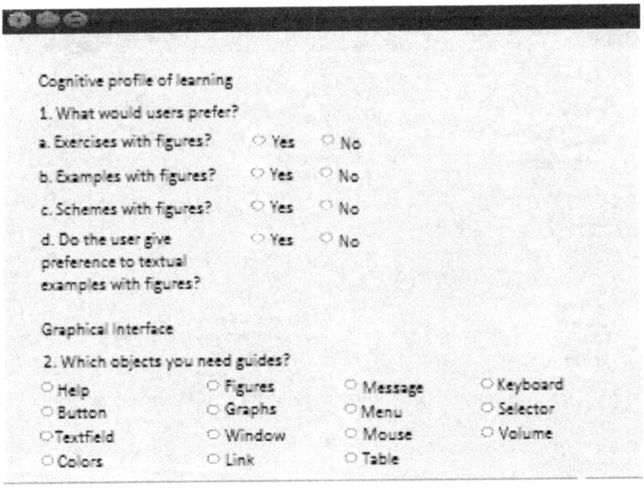

Figure 7. Cognitive Learling Style elicitation.

As example is shown in Figure 5, the rule knowledge base for people with ADD related to

children and elderly people. It was increased to the knowledge base tool.

For the construction of the selection rules we cross information of users of ECAs with ADD and ADHD disorders and other characteristics, we used the parameter age (child and adult). The resulting guidelines for the rule R1, for example, is shown in the Figure 6, which was selected set of guidelines for users with ADD and set of guidelines for user-child.

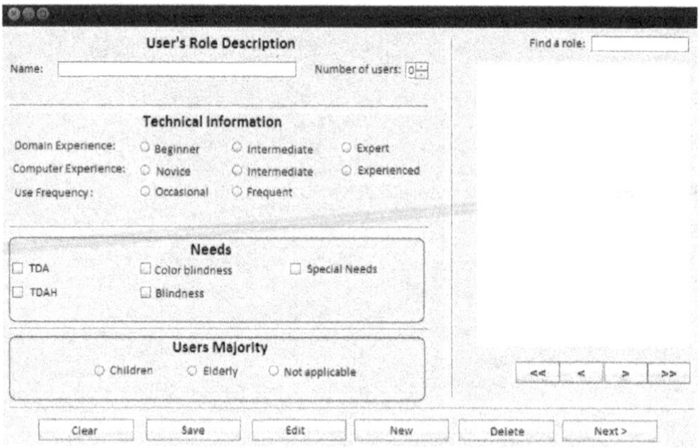

Figure 8. User Profile Analysis

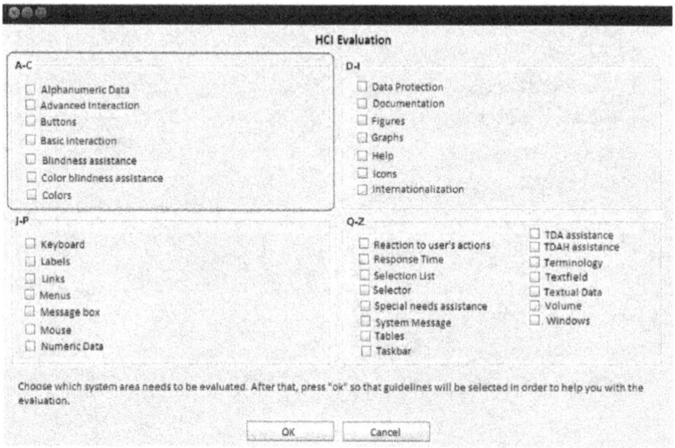

Figure 9. HCI evaluation interface

We recommend changes to the GuideExpert interfaces because of the addition of new taxonomies. The changes were suggested in the items: task analysis (because it does not allow the user to choose the user "child" and also to choose the needs of the users); context analysis (new items of graphical user interface were added); evaluation of interface design (new items of choices were added for the visual deficient, special needs, ADD, ADHD and others).

GuideExpert user interface was updated with new questions related to the cognitive learning style, as shown in Figure 7.

Figure 8 shows the elicitation of the difficulties. The question is if the user has some kind of need such as: ADD, ADHD, color blindness, visual impairment, other disabilities. The system also questions about the user's age (Figure 8, highlighted in red).

The other GuideExpert user interfaces that have changed refer to task analysis, where new items were added to the description of the GUI and interface evaluation. Options have been added to select among different profiles: child users, visually impaired, special needs, ADD, ADHD (Figure 9, highlighted in red). The screen shown in Figure 9 allows the Guide Expert system to select a series of recommendations to evaluate an interface. In order to allow this, the user must highlight the main features of the interface.

By extending GuideExpert it will be possible to specialize more and more recommendations; it will help the designer to automate a way of selecting guidelines that will guide the design or evaluation of interfaces.

4. Conclusions

It was observed during this study through the references related to the proposed theme, authors are conceptualized as Nielsen, Shneiderman and Plaisant, making several recommendations for building interfaces for children, elderly, etc. However, most of the recommendations deals with isolated aspects of the characteristics of users. It was noticed a large gap in this area in order to relate more than one feature. Given this problem, this study examined the learning styles and attentional deficits, allowing to generate a series of recommendations, guidelines, that fit the specific characteristics of the users profile. At the stage of acquisition of this new knowledge it was used as basis the class ontological description related to the content of the knowledge of GuideExpert. Following the methodology for ontology construction, it was made the acquisition of knowledge and conceptualization of new classes. Thus, a new taxonomy was added to the GuideExpert system together with the guidelines. The use of these recommendations helps the designer to interface with more knowledge giving the possibility to access them in an automated fashion and with various features, resulting in better recommendations and with best models specified by users.

Author details

Sandra Rodrigues Sarro Boarati, Cecilia Sosa Arias Peixoto and Cleberson Eugenio Forte

Methodist University of Piracicaba – UNIMEP, Brazil

References

[1] Amaral, A., & Guerreiro, M. (2001). Transtorno de déficit de atenção e hiperativi-
     dade: proposta de avaliação neuropsicológica para diagnóstico. *Arq Neuropsiquiatria*,
     59(4), 884-888.

[2] Angele, J., Fensel, D., Landes, D., & Studer, R. (1998). Developing Knowledge-Based
     Systems with MIKE. *Journal Automated Software Engineering*, Kluwer Academic Pub-
     lishers Hingham, MA, USA, 5(4), 389-418, doi> A:1008653328901.

[3] Bica, F., Souto, M. A. M., Vicari, R. M., Oliveira, J. P. M., de Zanella, R., Vier, G., Sou-
     za, K. B., Sonntag, A. A., Verdin, R., Madeira, M. J. P., Charczuk, S. B., & Barbosa, M.
     (2001). Metodologia de construção do material instrucional em um ambiente de ensi-
     no inteligente na web. *XII Simpósio Brasileiro de Informática na Educação SBIE- UFES*,
     21,23 November, Vitória, ES, Brazil., 1, 374-383, 8-58844-259-0.

[4] Breitman, K. (2006). Web Semântica: A Internet do Futuro. 8-52161-466-7, Rio de Ja-
     neiro.

[5] Chimaera. (2012). http://www.ksl.stanford.edu/software/chimaera, accessed 25 April.

[6] Cinto, T., & Peixoto, C. (2010). Guidelines de Projeto de Interfaces Homem-Computa-
     dor: Estudo, Proposta de Seleção e Aplicação em Desenvolvimentos Ágeis de Soft-
     ware. *Relatório Científico PIBIC/FAPIC. UNIMEP*, Piracicaba. Brazil.

[7] Rosa Neto, F., & Poeta, L. S. (2006). Prevalência de escolares com indicadores de
     transtorno do déficit de atenção e hiperatividade (TDAH). *Temas sobre desenvolvimen-
     to*, 14(83), 57-62.

[8] Gennari, J., Musen, M., Fergerson, R., Grosso, W., Crubezy, M., Eriksson, H., Noy,
     N., & Tu, S. (2002). The evolution of Protégé: an environment for Knowledge-Based
     Systems Development. *International Journal of Human-Computer Studies*, 58, 89-123.

[9] Gruber, T. (1993). A Translation Approach to Portable Ontology Specifications.
     Knowledge Acquisition, 5(2), 199-220.

[10] Hickman, F., Killin, J., Land, L., Mulhall, T., Porter, D., & Taylon, R. (2002). Analysis
     for Knowledge-Based system: a practical Guide to the KADS Methodology. London:
     Ellis Horwood.

[11] Horrocks, I., Sattler, U., & Tobies, S. (1999). Practical Reasoning for Expressive De-
     scription Logics. in proc of the 6th int. conf. on logic for programming and automat-
     ed reasoning (LPAR 99). September. Springer-Verlag. H. Ganzinger, D. McAllester
     and A. Voronkov (eds.), 161-180.

[12] Nielsen, J. (2010). Children's Websites: Usability Issues in Designing for Kids. Alert-
     box: september 13, http://www.useit.comalertbox/children.html.

[13] Nielsen, J. (1993). Usability Engineering. *Boston: Academic Press.*, 0-12518-405-0.

[14] Noy, N. F., & McGuinness, D. L. (2001). Ontology Development 101: A Guide to Creating Your First Ontology. *Relatório Técnico, Stanford University, Stanford.,* http:// www.ksl.stanford.edu/people/dlm/papers/ontoloy101/ontogy101-noymcguinners. html, accessed 20 January 2012.

[15] Oliveira Netto, A. A. de. (2004). IHC: Modelagem e Gerência de Interfaces com o Usuário. *Florianópolis: Visual Books,* 85-7502-138-9.

[16] Ontolingua. (2012). Disponível em:, http://www.ksl.stanford.edu/software/ontolingua, accessed 25 April 2012.

[17] Protégé. (2012). http://protege.stanford.edu, accessed 10 April 2012.

[18] Rosa, Neto. F., & Poeta, L. (2006). Prevalência de escolares com indicadores de transtorno do déficit de atenção e hiperatividade (TDAH). *Temas sobre desenvolvimento,* 14(83), 57-62.

[19] Russel, S., & Norvig, P. (2003). Artificial Intelligence, (2nd edition). *New Jersey: Prentice Hall Series.,* 0-13790-395-2.

[20] Shneiderman, B., & Plaisant, C. (2009). Designing the User Interface: Strategies for Effective Human-Computer Interaction, (5th edition). Reading, MA: Addison-Wesley Publishing Co, 0-32153-735-1.

[21] Souto, M. A. M. (2003). Diagnóstico on-line do estilo cognitivo de aprendizagem do aluno em um ambiente adaptativo de ensino e aprendizagem na web: uma abordagem empírica baseada na sua trajetória de aprendizagem. PhD thesis. Universidade Federal do Rio Grande do Sul, Porto Alegre.Brazil.

[22] Siqueira, C., & Gurgel-Giannetti, J. (2011). Mau desempenho escolar: uma visão atual. *Revista da Associação Medica Brasileira.,* 57(1), 78-87, 0104-4230.

[23] Uschold, M. (1996). Building Ontologies: Towards a Unified Methodology. *Proceeding of the Sixteenth Annual Conference of the British Computer Society Specialist Group on Expert Systems.,* 16-18, 16-18 December, Cambridge, UK.

Permissions

The contributors of this book come from diverse backgrounds, making this book a truly international effort. This book will bring forth new frontiers with its revolutionizing research information and detailed analysis of the nascent developments around the world.

We would like to thank Professor Eng. Petrică Vizureanu, Ph.D., for lending his expertise to make the book truly unique. He has played a crucial role in the development of this book. Without his invaluable contribution this book wouldn't have been possible. He has made vital efforts to compile up to date information on the varied aspects of this subject to make this book a valuable addition to the collection of many professionals and students.

This book was conceptualized with the vision of imparting up-to-date information and advanced data in this field. To ensure the same, a matchless editorial board was set up. Every individual on the board went through rigorous rounds of assessment to prove their worth. After which they invested a large part of their time researching and compiling the most relevant data for our readers. Conferences and sessions were held from time to time between the editorial board and the contributing authors to present the data in the most comprehensible form. The editorial team has worked tirelessly to provide valuable and valid information to help people across the globe.

Every chapter published in this book has been scrutinized by our experts. Their significance has been extensively debated. The topics covered herein carry significant findings which will fuel the growth of the discipline. They may even be implemented as practical applications or may be referred to as a beginning point for another development. Chapters in this book were first published by InTech; hereby published with permission under the Creative Commons Attribution License or equivalent.

The editorial board has been involved in producing this book since its inception. They have spent rigorous hours researching and exploring the diverse topics which have resulted in the successful publishing of this book. They have passed on their knowledge of decades through this book. To expedite this challenging task, the publisher supported the team at every step. A small team of assistant editors was also appointed to further simplify the editing procedure and attain best results for the readers.

Our editorial team has been hand-picked from every corner of the world. Their multi-ethnicity adds dynamic inputs to the discussions which result in innovative outcomes. These outcomes are then further discussed with the researchers and contributors who give their valuable feedback and opinion regarding the same. The feedback is then collaborated with the researches and they are edited in a comprehensive manner to aid the understanding of the subject.

Apart from the editorial board, the designing team has also invested a significant amount of their time in understanding the subject and creating the most relevant covers. They scrutinized every image to scout for the most suitable representation of the subject and create an appropriate cover for the book.

The publishing team has been involved in this book since its early stages. They were actively engaged in every process, be it collecting the data, connecting with the contributors or procuring relevant information. The team has been an ardent support to the editorial, designing and production team. Their endless efforts to recruit the best for this project, has resulted in the accomplishment of this book. They are a veteran in the field of academics and their pool of knowledge is as vast as their experience in printing. Their expertise and guidance has proved useful at every step. Their uncompromising quality standards have made this book an exceptional effort. Their encouragement from time to time has been an inspiration for everyone.

The publisher and the editorial board hope that this book will prove to be a valuable piece of knowledge for researchers, students, practitioners and scholars across the globe.

List of Contributors

João Inácio Da Silva Filho, Alexandre Shozo Onuki, Luís Fernando Pompeo Ferrara, Maurício Conceição Mário, Dorotéa Vilanova Garcia, and Alexandre Rocco
UNISANTA - Santa Cecília University, Brazil

José de Melo Camargo and Marcos Rosa dos Santos
AES - Eletropaulo Metropolitano Eletricidade de São Paulo S.A, Brazil

Ivan N. da Silva and Rogério A. Flauzino, Ricardo A. S. Fernandes and Danilo H. Spatti
University of São Paulo (USP), Brazil

Carlos G. Gonzales,Paulo G. da Silva Junior and Erasmo S. Neto
São Paulo State Electric Power Transmission Company (CTEEP), Brazil

José A. C. Ulson
São Paulo State University (UNESP), Brazil

L. G. Kabari
Department of Computer Science, Rivers State Polytechnic, Bori, Nigeria

E. O. Nwachukwu
Department of Computer Science, University of Port Harcourt, Port Harcourt, Nigeria

José Luis Calvo Rolle
University of Coruña, Spain

Héctor Quintián Pardo
University of Salamanca, Spain

Sandra Rodrigues Sarro Boarati, Cecilia Sosa Arias Peixoto and Cleberson Eugenio Forte
Methodist University of Piracicaba – UNIMEP, Brazil